Fmc 4J

THE NEW INNOVATORS

How Canadians Are Shaping the Knowledge-Based Economy

Roger Voyer and Patti Ryan

James Lorimer & Company, Publishers
Toronto, 1994

©1994 by Roger Voyer and Patti Ryan

All rights reserved. No part of this book may be reproduced or transmitted in any form or by any means, electronic or mechanical, including photocopying, or by any information storage or retrieval system, without permission in writing from the publisher.

James Lorimer & Company Ltd. acknowledges with thanks the support of the Canada Council, the Ontario Arts Council and the Ontario Publishing Centre in the development of writing and publishing in Canada.

Canadian Cataloguing in Publication Data

Voyer, Roger, 1938-

The new innovators: how Canadians are shaping the knowledge-based economy

Includes bibliographical references and index.

ISBN 1-55028-463-0 (bound)

1. High technology industries - Canada. 2. Information technology - Economic aspects - Canada. I. Ryan, Patti.
II. Title.
HC120.H53v68 1994   338.4'762'000971   C94-931732-2

James Lorimer & Company Ltd., Publishers
35 Britain Street
Toronto, Ontario M5A 1R7

Printed and bound in Canada

Contents

Acknowledgements	v
1 From Invention to Innovation	1
2 Technological Innovation	6

The Start-Up Firms

3 Vortek Industries Ltd.: A Solution in Search of a Problem	20
4 Quadra Logic Technologies Inc.: The Rough and Tumble of Focusing on a Direction	32
5 Xillix Technologies Corp.: Matching an Opportunity with Seasoned Management	45
6 Techware Systems: Techies Doing Their Own Thing	57
7 Instantel Inc.: Getting a Second Wind	70

The Growth Firms

8 Creo Products Inc.: A Seasoned Technology Manager Takes On a Mid-Life Challenge	84
9 DY 4 Systems Inc.: Four Dynamic Engineers in Search of a Product	95
10 Corel Corp.: Entrepreneurship, Experience and Deep Pockets	108
11 Newbridge Networks Corp.: From Near Disaster to Success	121
12 Dynapro Systems Inc.: Steady Growth through Alliances and Acquisitions	137

The Mature Firms

13 Cognos Inc.: Balancing Entrepreneurship and Professional Management	151
14 Mitel Corp.: Renewal after Rapid Growth and Turbulence	169
15 Gandalf Technologies Inc.: Rationalizing Too Many Products	186
16 International Submarine Engineering: From the Depths of the Oceans to Outer Space	201
17 Coping with the S Curve	214
18 The Context for Technological Innovation in Canada	223
List of Acronyms	231
Bibliography	233
Index	235

Acknowledgements

The idea for this book came to me while I was putting together material for a course on technological innovation that I was teaching at Carleton University. Material on the experiences of Canadian firms was very scarce, so I decided to develop some. I discussed the idea of writing a book for the educational market on the dynamics of technological innovation in Canada with Jim Lorimer. He encouraged me to do so and was instrumental in obtaining a grant from the Ontario Arts Council so that I could hire a researcher for the project. For this I am very thankful because, otherwise, the project would not have gone ahead and I would not have met Patti Ryan.

Patti, a graduate of the School of Journalism at Carleton University, did all the interviews and wrote up the case studies for this book. She did an excellent job. Her tenacity in getting to the right people within the firms has meant that we were able to present the views of very senior people and even founders of firms. Their insights on the ups and downs of bringing their ideas to market make this book come alive. For this I would like to thank Patti very much.

I would also like to thank the representatives of the firms who agreed to be interviewed, and hope that others can learn from their experience and insights into the innovation process.

I would also like to recognize Peeter Kruus, the Chair of the Technology, Society and Environment Studies program at Carleton University, who invited me to give a course on innovation to fourth-year students. It has been a very rewarding experience.

Roger Voyer
Ottawa, Ontario
September 1994

1

From Invention to Innovation

The purpose of this book is to demystify the process of technological innovation. While many people believe that innovation is synonymous with a flash of genius, there are underlying principles that may be set out. This book will outline these principles and will show how Canadian innovators are shaping the new knowledge-intensive industries.

What is technological innovation? Basically, it is the ability to transform an idea into a marketable product, process or service. Traditionally Canadians have been very good inventors or generators of ideas. The light bulb, the telephone, the zipper, the snowblower, the snowmobile, instant mashed potatoes, the Jolly Jumper baby seat, the paint roller, the table top hockey game, IMAX film projection and AM radio are all Canadian ideas.

Some of these inventions have been transformed into technological innovations in Canada and have launched successful companies. For example, the snowmobile is at the root of the success of the Quebec firm, Bombardier, and communications satellites are linked to the formation of Telesat Canada.

But many Canadian ideas, possibly too many, have been commercialized by others, meaning that the major benefits did not accrue to Canada. After Henry Woodward of Toronto was awarded a patent in 1875 for his incandescent light bulb, he sold a share of the patent to Thomas Edison because he could not find the money to commercialize the idea in Canada. This phenomenon of successfully inventing but failing to capitalize has been repeated many times in Canadian history.

In his book, *Ideas in Exile* (McClelland and Stewart; 1967), J.J. Brown documented the history of Canadian inventions from 1500 to 1966. It is a rich history, especially in the nineteenth and twentieth centuries. During this period of rapid industrialization, Canadians were responsible for major inventions in every sphere of economic activity. However, Brown's analysis led him to conclude: "The paradox that enlivens the history of Canadian invention is that Canada is a great producer of ideas, yet it has virtually no native technical industries."

Canada's historical focus on the exploitation of natural resources did not encourage the type of innovation that would create technology-based industries. The increasing globalization of the world economy and rapid technological change in the years since Brown wrote his book, as well as a better understanding of the innovation process and the resulting benefits, are forcing Canadians to rethink existing economic strategies so that inventions are transformed into marketable innovations within Canada. The historical strategy of importing technology and equipment to exploit our natural resources so that we can sell them in foreign markets has served us well until now. However this strategy is reaching its limits.

Canadians have sensed this and new innovators are emerging to capture the growth industries of tomorrow — knowledge-intensive industries such as computers, telecommunications, aerospace, instrumentation, chemicals and pharmaceuticals. These sectors have been capturing an increasing proportion of world trade: more than 25 percent today, up from 10 percent in 1980. Many of these new technologies can be applied to upgrade Canada's traditional natural resources sector.

Current Dynamics

In this century, technological innovation originated principally from well-organized research and development (R&D) activities within firms. In fact, Thomas Edison, the ultimate entrepreneur, was a major influence on the institutionalization of R&D in laboratories through the trial and error process that he created to develop inventions. This approach has changed dramatically in recent years. For example, Gandalf, an Ontario company that develops communications network equipment, frequently questions whether it makes more sense to make or to buy new technology. If the market in which the technology is to be applied is only developing slowly, deciding to make it would be sensible because the development of proprietary technology could provide a competitive edge. But if the area is developing rapidly, as is now increasingly the case, then a decision to buy might make more sense because the timing in bringing a product to market could be critical. A decision to buy can take many forms — licensing the needed technology, entering into a partnership with a knowledgeable firm or research organization to develop the technology needed rapidly, or even acquiring a firm with the technology.

This type of strategy, which reaches out beyond the firm, has come to be called search and develop (S&D) in contrast to the traditional R&D strategy. S&D has long been a pillar of Japanese industrial strategy and the industrial success of Japan in the last 20 years is well known. For example, while the VCR was invented by Ampex, a US firm, and the CD player by Philips, a Dutch firm, both were commercialized by Sony of Japan. With increasingly rapid technological change and the Japanese showing the way, S&D has become the strategy of choice of firms around the world.

Governments support the strategies of the firms in their jurisdictions through research and development programs and tax incentives because of the belief that technological innovation leads to economic activity, job creation and the socio-economic well-being of their citizenry, not to mention the taxable revenues that can be derived. This belief is supported by economic analysis, such as that of Robert Solow of MIT, who won the 1987 Nobel Prize in economics for demonstrating that technology is just as important as other factors, such as capital, in determining a country's level of economic growth.

The Determinants of Technological Innovation

What determines the ability of a firm to innovate and therefore survive in global markets are firm size, access to capital, ownership and government support.

While large firms have the technical and financial resources to ensure a continued stream of innovations, they also tend to be bureaucratic, which can inhibit innovation. For example, IBM's commitment to the mainframe computer, which is at the core of its business, has also been its Achilles heel; the company downplayed the emergence of smaller powerful computers, and its sales have suffered. This has led to a reorganization of the firm into a federation of smaller autonomous units to stimulate innovation. Large firms committed to R&D do invent but they can have difficulty translating these inventions into marketable innovations. Their presence in the market with a family of well-accepted products reinforces their commitment to existing innovations.

The survival of small entrepreneurial firms depends on their ability to bring innovative products to the market place. These firms can make inroads against well-established firms through innovation. For example, many computer firms challenged IBM with a variety of smaller, cheaper, more powerful computers. However, small firms do not have the resources to offer a broad range of related products and they usually survive by "betting the firm" on one product line. If successful, then they can begin to offer a wider range of products and, possibly, related services.

While large firms usually have little difficulty obtaining the capital needed for growth because of their assets and track record, small firms can have great difficulty. Small firms are selling a promise and as such can have difficulty convincing investors of the merits of their ventures. It is not unusual for entrepreneurs to mortgage their homes and to borrow money from family and friends to launch their firms. If the launch is successful, it becomes easier to obtain funds from the financial community (for example, venture capital or a bank loan).

The question of ownership is important because R&D is a corporate function and is usually undertaken within the parent firm. Subsidiaries in other countries usually obtain the innovations from the parent. This means that the subsidiary's role is to adapt specific innovations to its needs. Of course, the subsidiary has to pay the parent for the technology that it gets. The potential for true technological innovation is much less than if the subsidiary undertook its own R&D and related commercialization functions on site. Canada has a particularly high level of foreign ownership. In manufacturing, for example, Canada, along with Australia and Ireland, has more than 30 percent of its firms in foreign hands, well above the level in other OECD countries. In technology-intensive sectors, foreign ownership is especially high. In 1991, foreign control of Canada's electrical and electronics firms was 66 percent; in machinery and equipment it was 43 percent; and in transportation it was 58 percent.

In recent years, larger firms have been establishing R&D facilities around the world to capture the technical skills available in various regions. This process is not limited to industrialized countries. For example, firms have been setting up software R&D facilities in India.

The financial and infrastructure support provided by governments is an important determinant of technological innovation. Governments provide not only tax incentives, grants and contracts to stimulate innovation, they also provide the infrastructure — that is, the universities and colleges, publicly funded laboratories, advanced communications (for example, high-speed networks) and transportation systems (for example, France's high-speed TGV train system) that are needed to support the development of innovative firms.

In Canada, because of our history of natural resource exploitation, we have many large firms in the resource sectors. However, these firms do not undertake much R&D, and their technology strategy is to adapt foreign technology to permit more efficient exploitation of natural resources. Moreover, many companies exploiting natural resources in Canada are foreign-owned, which means that they largely depend on their parent company for technology.

In the advanced technology sectors Canada has very few large firms. However, these firms, such as Northern Telecom, do undertake sizable R&D programs through which new product innovations emerge. These firms strongly rely on a product innovation strategy.

However, the majority of Canadian firms are small or medium-sized. Of the one million or so firms operating in Canada, some 97 percent have fewer than 50 employees. These firms usually have major difficulties in obtaining financing because of their size and limited track record and because Canada's financial community continues to be oriented towards investment in the traditional sectors rather than the emerging technology-intensive sectors. A cultural change is very much needed in the financial community.

The same cultural change also needs to occur within governments in Canada. The limited available government support needs to be oriented towards "sunrise" industries (such as electronics, software, telecommunications, medical devices) rather than "sunset" industries (such as resource extraction). The sunset industries should be encouraged to develop value-added products and to establish close relationships with Canadian suppliers to create new products and processes.

The factors affecting technological innovation are highlighted in the following pages through the presentation of case studies of innovative Canadian firms. To ensure full appreciation of the innovation taking place within these firms, some basic concepts related to technological innovation are presented in the next chapter.

2

Technological Innovation

In discussions of technological innovation certain words are often used interchangeably. It is important at the outset to set down some working definitions of the key concepts that will be presented later in this chapter. The key words are:

- *discovery*, which is the revelation of what was previously not known;
- *invention*, which is something created through thought;
- *technology*, which is practical knowledge; and
- *innovation*, which in this context is the process of commercializing an invention or technology.

Scientific discovery cannot be patented, but invention can. In recent years, however, the line between discovery and invention has become increasingly blurred. In the field of genetics, for example, once a human gene is separated by some technique from the cell in which it is found, it can be patented because it is no longer part of a natural system; it becomes an "invention." Understandably, much controversy surrounds the possible commercialization of what could be considered the discovery of new knowledge.

Invention and technology have always been part of the innovation process. For example, the invention of the potter's wheel in Egypt before 4000 B.C. led to the innovation of cooking pots; in the Middle Ages, the invention of the magnifying glass led to eyeglasses; and the invention of the steam engine in 1769 led to innovations such as the locomotive, the steamship and various industrial processes. The invention of the transistor in 1940 initiated the microchip revolution. Now discovery is also becoming an element of that process. Since the emerging knowledge-based industries are very much science-based, it is to be expected that discovery will make an increasingly important contribution to innovation.

The Innovation Process

Basically, the innovation process is made up of the steps needed to bring an idea to the market place. These steps are shown, in their simplest and most idealized form, in Diagram 2-1.

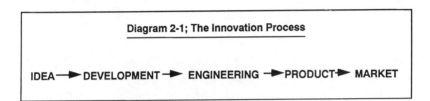

Diagram 2-1; The Innovation Process

IDEA ➤ DEVELOPMENT ➤ ENGINEERING ➤ PRODUCT ➤ MARKET

While these steps are shown as sequential, this is not necessarily the way innovation occurs. For example, the arrows could flow the other way; the idea could come from the market place — from a real need — rather than from intuition or scientific research. This would be called market-pull innovation as opposed to the technology-push innovation shown in Diagram 2-1. In fact, an innovation can be generated at any stage of the innovation process.

Innovation builds on the international pool of past scientific discoveries and inventions. This is possibly best illustrated by the chronology of key events leading to the development of the electronic computer:

Date (circa)	Event
1700	A German mathematician, Leibniz, discovers that all numbers can be represented in a binary system of 0 & 1;
1850	The English inventor, Babbage, invents the mechanical calculator which presaged the development of the modern digital computer;
1910	Two English logicians, Whitehead and Russell, link logic and mathematics;

1915	The American statistician, Neurath, determines that all information can be quantified;
1915	An American scientist, DeForest, invents a vacuum tube to convert electrical impulses to sound waves (the audion tube);
1935	IBM engineers develop the audion tube into a binary switch; and
World War II	The war catalyzes the development of the first electronic computer, ENIAC.

Various discoveries and inventions from all over the world, as well as the large sums of money spent on computer development during the war, had to come together to bring about this innovation. In fact, the electronic computer only became a reality after World War II. If any one of the key events had been missing, this innovation would not have come about at that time.

Innovation can lead to new and improved products (such as computers), to processes (such as the use of robots on production lines), and to services (such as new approaches to teaching). Usually, innovation is incremental; existing products, processes and services undergo improvements. However, periodically an innovation brings about a profound change, such as the introduction of the personal computer, which has led to completely new ways of working.

Innovation also contributes to the way we organize major aspects of society. For example, the co-operative university program pioneered at the University of Waterloo, where students spend time in a work environment as well as at the university, is an important innovation. The Canadian health care system is also a major innovation.

However, in this book the focus will remain on technological innovation leading principally to the development of new or improved products.

The S Curve

The innovation process described above is not linear. It can be broken down into at least four stages, which taken together usually follow the shape of an S curve, as shown in Diagram 2-2, when the level of effort to commercialize a product (using such measures as cumulative R&D expenditures and cumulative person-years) is plotted against an indicator of performance of the product (measured by sales or a technical improvement parameter). Since effort takes time, time is sometimes used as a proxy for effort in constructing S curves.

The first stage of the innovation process is to transform the idea to a workable product prototype. Transformation takes a significant level of effort and money and is very risky. This is the flat part at the bottom of the S curve. This activity sets the basic performance parameters for the product. This is also the expensive part of the development process. A general guide that is often applied is that if it costs $1 to develop the idea, it will cost $10 to engineer it and $100 to produce it!

The second stage occurs when the product is first introduced into the market. The first sales are made, "bugs" are worked out of the product and, if successful, sales begin to multiply. The product begins to climb the S curve. There can be a lot of overlap between the

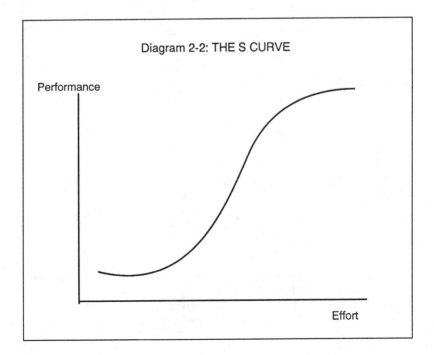

Diagram 2-2: THE S CURVE

first and second stages as approaches to product development and marketing are refined.

The third stage is the rapid growth part of the S curve where the product is gaining market acceptance and revenues accrue with less effort. This is often regarded as the pay-off stage.

The fourth stage occurs when the market becomes saturated or technological improvements are exhausted. Performance slows down and might even decline. At this stage a lot of effort is needed to obtain even small improvements in performance. However, at times a totally new product emerges to better fill the same need as the exiting product. If this occurs, then a new innovation process, resulting in a completely new S curve, will be set in motion. An example of the jump from one S curve to another with ever better performance is shown in Diagram 2-3.

What Diagram 2-3 indicates is that, with some effort, rayon displaced cotton in tire cord, and then nylon took over from rayon, with

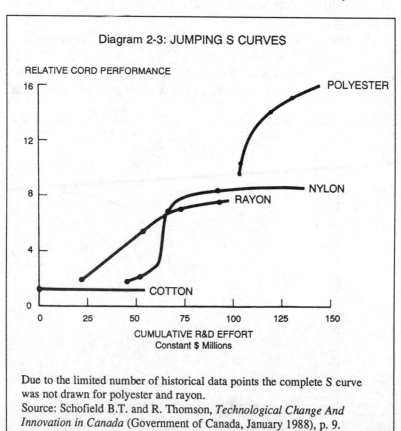

Due to the limited number of historical data points the complete S curve was not drawn for polyester and rayon.
Source: Schofield B.T. and R. Thomson, *Technological Change And Innovation in Canada* (Government of Canada, January 1988), p. 9.

polyester eventually demonstrating such superior performance that it displaced nylon. But the level of effort continued to increase.

A "battle of S curves" has been taking place in everyone's living room. Some 90 percent of Canadian households have cable television, which brings in about 50 channels. However, some entrepreneurs have proposed that they can deliver some 500 channels directly to the home via satellites, by-passing the cable network. All that is required in the home is an affordable (less than $1000) pizza-sized dish antenna. The cable companies, which are at the top of their S curve, have fought back with a multi-billion-dollar scheme aimed at providing comparable services. The battle, as with all S curves, is between vested interest and innovation.

The S curve can describe the growth of a firm as well as the development of a product. In fact, the parallel is easy to understand because most firms begin with only one product. In firms, difficulties usually arise in the third and fourth stages of development. In the third (fast growth for least effort) stage the firm has to move from an entrepreneurial style to a more professional form of management to manage growth, to obtain the financing needed to keep growing and to develop the next generation of products.

Going from the level of the firm to the level of industrial sectors, S curves can also be detected as shown in Diagram 2-4. Entire industries grow, peak and decline as they are replaced by new industries. And as shown in Diagram 2-4 we are now entering into a new knowledge-based industrial age.

The same can also be seen at the societal level. In the last century, Britain developed rapidly due to the industrial revolution. In this century, the United States replaced Britain as the world's industrial engine. And it now looks as if Japan and countries of Southeast Asia are in the ascendancy.

The S curve is a useful tool to describe and analyze the development of products, firms, industries and even societies.

Intellectual Property

How do you protect an idea to ensure that someone else does not get it to the market place first? Governments have set in place mechanisms to enable people protect their ideas or, to use a technical term, their intellectual property (IP).

In Canada there are five major ways to protect IP: patents, copyrights, trademarks, industrial designs and trade secrets. Each ap-

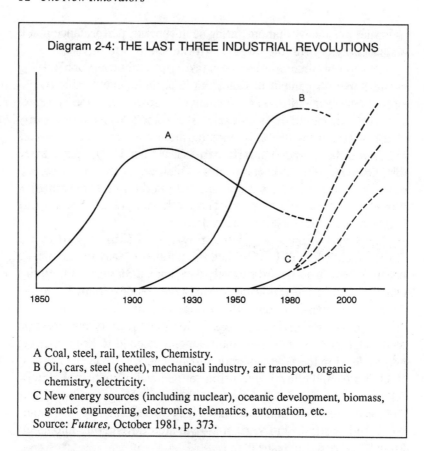

Diagram 2-4: THE LAST THREE INDUSTRIAL REVOLUTIONS

A Coal, steel, rail, textiles, Chemistry.
B Oil, cars, steel (sheet), mechanical industry, air transport, organic chemistry, electricity.
C New energy sources (including nuclear), oceanic development, biomass, genetic engineering, electronics, telematics, automation, etc.
Source: *Futures*, October 1981, p. 373.

proach is supported by legislation. A fifth approach is trade secrets, which is supported by Canadian common law.

Patents provide exclusive rights for the exploitation of an idea to the inventor for a limited time. In Canada that exclusive period is 20 years. The Canadian Patent Office uses three criteria to determine if a patent should be granted: novelty, utility and inventive ingenuity. Once a patent has been granted, it is made public so that the information is shared with the world while protecting the inventor's exploitation monopoly.

Copyrights are awarded to authors and creators of original works. These include books, music, films and works of art. Copyrights protect against the unauthorized uses of original works, usually for the lifetime of the creator plus 50 years.

Trademarks are visible symbols that distinguish the products or services of one firm from another. A trademark is registered for a period of 15 years, but can be renewed indefinitely.

Industrial designs protect the designer of products against unauthorized uses for a period of 10 years.

Trade secrets are confidential, commercially valuable information. For example, the formula for Coca-Cola is a trade secret. Breaking the confidence of a trade secret can be punished by law.

In technology-intensive industries, with short product cycles, there is a growing preference for trademarks and trade secrets over patents and copyrights. The latter are expensive to obtain and do not guarantee that a legally protected idea will not be superseded in a world where technology is changing so rapidly. Increasingly, firms depend on know-how that will help them keep ahead of the competition.

Technology Transfer

Technology transfer is the transfer of technical knowledge from the source of that knowledge to someone wanting to use it. This transfer can occur either within or from outside the firm.

The traditional approach to technology transfer within a firm is from the R&D laboratory to the production division. These two units do not necessarily have to be housed on the same site. They can even be in different countries. One of the fastest means to transfer technology is from the parent firm to a subsidiary. However, more recently, teams of researchers and production engineers have been working together to carry an idea from the development stage to production. In this way the technology transfer process is ongoing in a fully integrated manner. This approach has been found to be very effective.

Increasingly technology must come from outside the firm. Following S&D strategies, firms obtain technology from a variety of sources, including licensing patents, acquiring other firms, getting know-how from universities and government laboratories, entering into R&D partnerships with other firms and research institutions and hiring people with specialized knowledge.

Other linkages that come into play in fostering technology transfer include producer-user relationships, contractor-subcontractor links, consultant-firm relations and various informal processes, such as seminars, workshops and even cocktail parties.

Diagram 2-5: TECHNOLOGY-MARKET MATRIX

	OLD Technology	NEW Technology
NEW Market	"Fad" Markets eg. Hula hoop No market stability	No market history and high technology risks eg. Video home shopping Good profit margins but very high risk because the market may not accept the product
OLD Market	Proven Markets and Technology eg. Appliances Low profit margin	Some market history available eg. electronic thermostats, portable field instruments Good profit margins

Source: Doyle D., *Making Technology Happen*, 3rd. ed. (Doyletech Corp., 1992).

Market Analysis

Having a good grasp of the market for the products, processes or services based on the technology to be exploited is essential. As shown in Diagram 2-5, there can be four categories depending on the maturity of both the technology and the market. While the most promising technology is new technology, the most promising market is an old market where a discontinuity in the S curve can be jumped. For example, it appears that new satellite communications technology will permit direct broadcast to the home, by-passing the now-accepted cable network route. However, the market information that has been gathered over the years regarding cable distribution will be of use in estimating the market penetration of the new technology.

Market research is needed to determine the size of a market and the rate of penetration that can be expected. This information is then used to determine the revenues that can be expected. This information is fed into a business plan as described below.

The Business Plan

All of the above concepts come together in a business plan for a new company or a new venture within an existing firm. A business plan must present the business opportunity (i.e., the idea), the product(s) to be offered, the market to be penetrated, the costs of launching the venture, the investment required and the expected return on that investment. That is, all the factors that drive the innovation process must be defined and documented convincingly so that a decision to, or not to, invest in the venture can be made.

For example, in structuring a business plan, there has to be some assurance that the intellectual property is well protected if this is needed and that there is a product development strategy that will permit the firm to climb the S curve. One way to climb the S curve is to have a product migration strategy, that is, as shown in Diagram 2-6, to have a number of products that can be derived from the initial product to serve different markets. These products are determined in

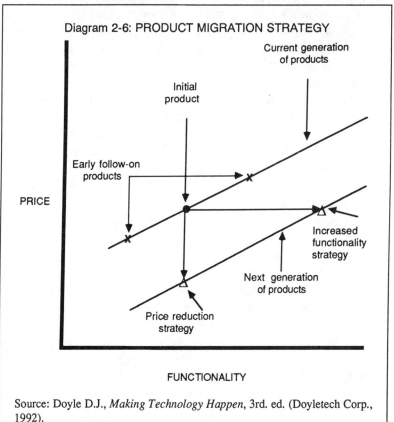

Source: Doyle D.J., *Making Technology Happen*, 3rd. ed. (Doyletech Corp., 1992).

terms of price and functionality. For example, differently priced automobile models have different functionalities (such as engine performance, number of seats and so on). These families of products are defined in terms of the technology that can be exploited and the markets that are to be served.

Financing Innovation

If a decision is made to develop an idea, that is, to set in motion a particular technological innovation process, financing has to be found. There are different sources of funds, depending on the nature of the venture.

A new venture, or "start-up," is usually difficult to finance because there is no track record. The entrepreneur usually has to finance the venture with the help of family and friends or knowledgeable individuals (who are sometimes referred to as "angels"). A professional venture capitalist can also advance funding, usually in exchange for equity participation in the venture. However, venture capitalists tend to wait until the venture has made some progress (that is, well along stage 1 and possibly into stage 2 of the S curve) before investing. Some funds can also be forthcoming from various government support programs.

As the venture progresses, and if it is successful, it will attract financing from venture capitalists, banks and eventually from the stock market. If the venture is launched within an existing firm, it can be easier to finance because the firm can provide internal resources, or if it already has an established track record it can probably get loans from financial institutions or put a new share offering on the stock market to obtain the needed capital.

Because it can take a long time to show a profit from a venture, possibly some five years or more, sizable investments are likely to be needed. The S curve can have a deep investment trough before initial sales are made, as shown in Diagram 2-7. This means that a variety of financial sources will have to come into play. Most opportunities flounder because of the lack of capital.

In Canada, it is especially difficult to get adequate financial support for technology-based ventures, principally because these ventures are relatively new and there is not much experience in financing them, especially in Canada's conservative banking community.

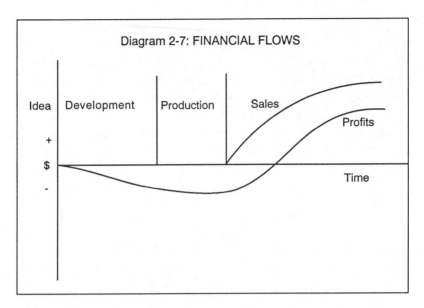

Managing Innovation

The approach to the management of innovation is linked to the position of the firm on the S curve.

At the idea stage, the entrepreneur is very much the manager in transforming the invention into a commercial reality. Since this is the developmental stage, the entrepreneur can be seen fulfilling many functions — engineer, promoter, fund raiser, administrator.

As the firm takes off the various functions are separated because of the work load. Initially, the representational/marketing functions are separated from the technical functions. Sometimes this separation can be difficult to achieve because the original entrepreneur may not be willing to let go of certain functions. Many entrepreneurs are technical people who do not fully value functions such as marketing and selling. Often at this stage in the development of the firm, if outside venture capitalists are involved, they may insist on a professional manager to protect their investment. People experienced in starting up firms are usually difficult to find, especially in Canada because we have not had much experience with launching technology-based firms.

As the firm grows and climbs the S curve, the entrepreneur or the start-up manager has to give way to a professional management team. The founders of the firm do not necessarily have to leave, but all the key managerial, financial, marketing and engineering functions now must be undertaken by professionals if the firm is to make a successful transition from the entrepreneurial to the professional mode. Many firms have difficulties making this transition because the founders refuse to give up or share control, which usually means that either they fail or their growth slows dramatically.

If a successful management transition is made, then the firm has a good chance of becoming a mature firm with a recognized position in the market place. As the firm reaches this market position, it becomes increasingly concerned about the competition and about

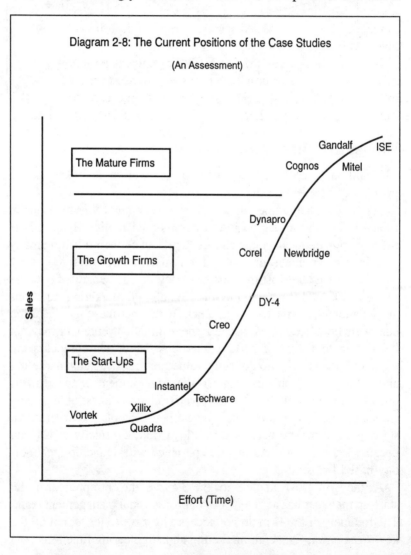

creating the new innovations that will ensure its presence in the market. It has to keep jumping to new S curves, because if it does not, others will.

Innovation in a large firm can become more difficult because functions become bureaucratized. It becomes a constant effort to ensure close links among research, development, manufacturing and marketing. Some large companies deliberately organize themselves into small units (say 100 people) to keep innovation alive. Others look outside and strike temporary alliances with other firms to develop technology that can be of mutual benefit. These networks of companies that come together to develop or exploit specific ideas have come to be known as virtual corporations.

The S curve is a powerful construct because it permits the integration of all the key concepts necessary to study the innovation process, a process that is illustrated through the case studies presented in the following chapters. The order of presentation is in three categories: start-ups, growth and mature firms, as shown in Diagram 2-8. Each firm is placed where it is currently perceived to be on the S curve.

3

Vortek Industries Ltd.: A Solution in Search of a Problem

Vortek is the story of a company with leading-edge technology that has made the Guinness Book of World Records. *But finding solid markets for the technology has remained elusive.*

Vortek Industries Ltd. is the quintessential example of technology bent on finding a market. The Vancouver-based private company produces what has been called "pure corneal dynamite" — the world's most powerful light, now sold for use in a variety of industrial applications like thermal testing, semi-conductor processing and solar simulation. The Vortek lamp is based on the principle of an 11-cm-long electric arc passing through argon, which gives off illumination as a byproduct of the arc's 12,000°C temperature. This temperature is almost twice that of the sun. The arc is contained within a quartz tube cooled by a thin film of water spinning down the inner wall between the quartz and the gas.

The brainchild of a set of University of British Columbia physics researchers, Vortek was born in 1975 when David Camm, a research engineer, co-invented the lamp with physics professor Roy Nodwell. Working in a physics lab on the UBC campus, Camm was struck with the idea of cooling the arc of the lamp by surrounding it with water constantly circulating in a vortex; earlier lamps had burned out quickly when the arc's temperature became too hot. Camm and Nodwell turned the invention into a commercial enterprise in 1976 with money from both the Federal Business Development Bank and a National Research Council (NRC) grant to the university. The NRC in Ottawa contributed $330,000 and the Federal Business Development Bank provided a $250,000 equity investment. Funds were granted on condition that the researchers start up a company to produce and sell the technology.

In 1978, Camm and Nodwell were joined by Gary Albach and Stephen Richards, all PhDs at UBC. They jointly formed a research team that worked to develop the arc lamp. When their company first began to do business in 1978, the original arc began selling at between $70,000 and $200,000; the price depended on the application. The lamp's big selling point, at the time, was that it could cut both capital and maintenance costs in large-scale applications.

It was in 1978 that Camm, Albach and the others were forced to confront the million-dollar question: What do you do when you've invented the world's brightest light, and no one wants to buy it? Their answer: Create your own market by developing other uses for it. Researchers at Vortek turned the light into a high-tech, super-powerful furnace and began trying to sell it to space-age manufacturing companies. Most recently, still determined to find a use for it as a source of light, they've been marketing it through demonstrations to show its merits as an aid for search-and-rescue teams trying to carry out emergency work at night.

High-profile exhibitions have long been a feature of Vortek's marketing techniques. While Expo 86 was being built in Vancouver, Vortek took the opportunity to demonstrate — and experiment with — its lamp by lighting up the Expo site on the north side of False Creek near downtown Vancouver while construction continued overnight. The 100,000-watt arc lamp, worth about $100,000 at the time, was mounted on a truck on a bridge around dusk. The tube was eight inches long, one inch in diameter, and was surrounded by a reflector of about three square feet. Vortek has also held demonstrations of its light in both Thunderbird Stadium at UBC and at Cypress Bowl, in both cases using a single lamp to light up the entire stadium.

But a mere exhibition or two has not been Vortek's only brush with fame. Sometime in the mid-80s, Vortek spoke to Walt Disney Studios and the producers of the movie *Dune* about providing a bright light for special effects. No deal was ever made, because Disney decided such lighting, regardless of who provided it, would be too expensive. "They were originally planning to film all the giant worms moving in daylight," says Camm, "and in the end, they had to rewrite the script and have it look like dusk every time a worm came out."

In 1990, *The David Letterman Show* invited Vortek to melt a car with a megawatt model of the lamp when they heard that it was capable of zapping a small car into slag. (Albach refused, saying it was unsafe.)

Not least of all, 1986 marked Vortek's first appearance in the *Guinness Book of World Records*. The record book says, "Of continuing burning sources, the most powerful is a 313 kilowatt high-pressure arc lamp of 1.2 million candle-power, completed by Vortek Industries, of Vancouver, B.C., Canada in March 1984."

History and Milestones

The president, Reg Allen, gives overall direction to the company. On a day-to-day basis, the firm is run primarily by technology vice-president David Camm and executive vice-president Gary Albach. Steve Richards, who was the third co-founder, left the company in 1992 to pursue a career with another high-tech company in Boston.

Albach is in charge of running the business, while Camm, who was doing post-doctoral work in astrophysics when the company first got off the ground, is still heavily involved in research. Because of their scientific backgrounds and the small size of the company, both still have a hand in each area. Camm says he got involved after facing a choice between becoming a professor, working as a researcher at a big American plasma lab, or getting in on the Vortek venture and remaining in Vancouver. It was no contest, he says. "If I was going to stay in Vancouver, I had to start my own company," he says. "Well, I like Vancouver."

> An attractive physical, cultural and social environment is a strong inducement for "footloose" skilled people.

The most notable thing about this company is that over the course of its history, its focus hasn't shifted from research to marketing or management. In fact, with hindsight, Camm admits they probably would have been more successful had they paid more attention to marketing their product earlier on. The company currently reinvests some 30 percent of its modest revenues into research and development. Camm says this level would probably drop if Vortek found a niche and began selling a product regularly. Still, as he puts it, "there's a certain amount of money you have to spend to keep up, and with us, it's more a dollar value than a percentage."

Camm says starting up a company seemed like a great idea initially because they were sure somebody would want their product. "We were naive," he says now. "We were sure that we were making the best light source in the world and we were going to sell it to stadiums." Originally, they wanted to sell it for use as a floodlight.

However, as he now concedes, they have yet to sell one as a light source.

The company has its roots in research at UBC, where physicist Roy Nodwell got the first funding — a PRAI (Project Research Applied to Industry) grant from the NRC. The purpose was to encourage commercialization of university research by taking a professor's idea and running a few years of experiments to test and prove the technical feasibility of the product. Nodwell hired Camm as a research engineer. They worked at the university for two years, hiring a machinist and another post-doctoral researcher, and invented the lamp. At the end of the two years, they left the university and went in search of a collaborating company — a final step required by the PRAI grant. They found Annatech, another Vancouver company. Annatech's president's strong advice was for the researchers to start a company.

"So we did," says Camm. "And by the end of the two years [in 1976], we were three graduate students running our own corporation."

Camm, Albach and Richards got support from friends and relatives in the form of donations "so we wouldn't starve," says Camm. Then the Federal Business Development Bank, which had just started a venture capital division, invested in Vortek. "We were one of their first equity investments," says Camm. After the company had been around a few months, other government research programs, such as IRAP [Industrial Research Assistance Program], started to kick in and lend support. "That's how we managed to make some product and start selling," says Camm.

> Finding financing from all possible sources is the major challenge for companies starting up.

They had a business plan that said they were going to break even in five years; they did it in three, and in five were starting to make money, at which point "we started to really pay ourselves," says Camm.

Vortek's first big sale came in 1981, to the federal government. "They were looking at a facility to do solar simulator testing," says Camm. "They had solar panels that were going to go on buildings and they had to be able to test them in order to be able to certify that they met certain standards for indoors and outdoors. So they were building a facility to test these things, indoors and outdoors. We got into a competition to build and deliver an indoor lamp system. It's still running — it's called the NSTF (National Solar Testing Facility) and it's in Toronto." Around this time the company finally started to make some serious money.

"We'd had a prototype lamp system that we'd built more or less to prove the lamp really worked," says Camm. "That one got sold to BC Research, who used it to try to remove ice — the idea was, you could shine the light on the ice, it would go right through the ice, hit the pavement, warm up a very thin layer, and then you could scrape it off quickly before it froze again. It was to test the feasibility of its use on airport runways, mostly. If the Toronto airport shuts down because of an ice storm, for example, it could lose between $400,000 and $500,000. So you'd only have to use it twice before it paid for itself." To date, no airport has made them an offer.

Camm says they still haven't given up on the idea of using the lamp as an actual light instead of to provide heat. "Every so often we go back and try again," he jokes. "We did a major study on using it at the BC stadium, in log-sorting areas, industrial loading areas. We once installed it on the side of the mountain and lit up the whole valley. There were demonstrations at the Expo sight, up on Mount Seymour, specialized lighting applications for movies. And we spent a few man-months of time reviewing the lighting market to see whether there's any potential in areas like search-and-rescue applications. We put together a demonstration for the Coast Guard." They're also considering it for use in helicopters for those missions.

As Camm puts it, and as David Letterman was quick to point out, the lamp "can melt anything. Rock, tungsten, you name it." This is one of the reasons Vortek might grudgingly have to forget about unearthing a use for the lamp as a source of light. Its heating powers are hard to ignore and much more widely applicable. This may allow Camm and Albach to create new and different markets for the same product.

"The nice thing about our lamp is that it's just like sunlight," says Camm. "Plants love it. They think they're outside. Unfortunately, sunlight is not a very efficient sort of light — it contains lots of invisible forms of light, such as infrared and ultraviolet. So the percentage of visible light is actually quite small." Per watt, that's not too efficient; there would have to be an extraordinarily good reason for a potential buyer to need such a high-powered, compact source of lighting. This is why Vortek has had to branch out and find other uses for its light in order to survive.

In 1980, Vortek created its first solar simulator heater and made its first sale — to NASA. By the early 1980s the company had made a major decision to switch from the sports stadium market to radiant heating, sunlight simulation and thermal testing. "The market [for the lamp as a light source] just wasn't big enough to justify the engineering," says Albach.

In 1987, Vortek delivered a $500,000, two-metre-high photon furnace for testing nosecones and wing edges to a US Air Force base. Vortek agreed to provide, install and train the technical staff to operate the furnace, located in the flight dynamics laboratory at the Wright-Patterson Air Force Base in Dayton, Ohio. In a process similar to using a magnifying glass to focus the sun's rays on a particular spot to start a fire, the furnace was designed to irradiate materials with a power beam of white light, using two ultra-powerful arc lamps focused on a single spot. The 600,000 watts of continuous electrical power for the lamps were drawn from the hydro grid in Dayton.

That same year, Vortek delivered smaller furnaces to a Boeing plant in Wichita, Kansas, and to another Boeing plant in Seattle, where it would be used in testing parts for a proposed hypersonic airliner. Following these sales, the company geared up to produce four units a month, all of which found buyers at prices ranging from US$150,000 to US$500,000.

By 1988, Vortek's main customers were still NASA, the US Air Force and their contractors, but the company was starting to find itself moving away from military customers. Vortek arc lamps began to be used to harden steel and foundry coatings in Europe and Japan. At the beginning of 1988, Vortek delivered what it called "the world's most powerful photon furnace" to the Flight Dynamics Laboratory at Wright-Patterson Air Force Base in Ohio. The base, which paid $750,000 for two of them, intended to use the arc lamps to irradiate aircraft and space vehicle parts to simulate the high temperatures they encounter in actual use.

By now, there were 20 people on staff at Vortek, most of them working in research and development. The company had a partner in Boston doing most of the manufacturing, marketing and distribution. All the company's products were being exported, many of them to provide heat treatments for the semi-conductor industry. Vortek lamps were now selling for anywhere between US$150,000 and US$750,000 to big American companies like Eaton Corp., which incorporated them into equipment to be resold to companies like IBM, which in turn used the equipment to manufacture semi-conductor chips.

The company's next major milestone was a US$1.1 million contract to deliver a new generation photon furnace to test the X-30 "space plane" being developed by NASA and the US Air Force. The X-30 project was a US$1.6 billion program aiming to deliver a prototype, that could fly 25 times the speed of sound and take off from an ordinary airport directly into orbit, by 1997.

Around this time, Camm and Albach had become aware that the major commercial application for their superlamp was not as a source of light, but rather as one of heat. Instead of the floodlight they'd envisioned 10 years earlier — which had found no takers — their product had now settled into different heat applications. The intense light from Vortek's lamps is a formidable source of heat; targets placed a hand's breadth from the 300-kilowatt lamp's 11-cm-long tube can reach 3,000°C, or about half the temperature of the sun's surface. The market now included semi-conductor manufacturers, who used the lamps to anneal silicon (a process designed to remove the crystalline defects that show up in silicon wafers during photolithography). Vortek also began supplying lamps directly to customers such as IBM, Siemens and Motorola when the Eaton Co., which had originally incorporated Vortek's lamps into optical ovens, discontinued that line.

> The intended application for a technology is often not the one that has market appeal.

In the summer of 1990, Vortek signed a contract with a German industrial research institute, Laser Zentrum Hannover, that performs practical research and development work for European companies. Albach says the company views the Vortek lamp as "an economical method for doing what it had wanted to do with some lasers."

The following year, Vortek began to delve into a different area, competing with lasers. They discovered an area in which the lamp could outperform lasers: thermal hardening of metals. Vortek applied this technique to agricultural plow blades, which traditionally compromise hardness for tensile strength to resist breakage. A laser could duplicate this work, but it would be much less efficient. "The essence of a laser is that it's a light of one wavelength, and therefore you can focus it to a pinpoint," explains Albach. "It's like using a pencil to work on the surface of a metal. You'd have to scan it line by line, which would take a long time. With the lamp, you could scan the whole thing in seconds."

The auto industry was something else Vortek targeted that year, suggesting its lamp could be used in transformation hardening to improve car transmissions by improving the parts' surface quality.

Major European car manufacturers also expressed an interest in lamp-treated parts.

In the spring of 1993, Vortek was again cited in the *Guinness Book of World Records* for developing the world's most powerful continuously burning light source, aimed for use in search and rescue and emergency response.

> World records provide great publicity that a marketing expert can seize upon.

Vortek still manufactures custom lamp systems and optical assemblies for research at high radiant power densities. They continue to sell ultra-power lamps for a variety of industrial applications, such as thermal testing, semi-conductor processing and solar simulation. Solar simulators are used worldwide for indoor testing and certification of solar products and by NASA for solar-powered research. For semi-conductor processing, Vortek supplies the US and European markets with 100 kW lamps and optics as sources in optical and radiation annealers, used for high temperature processing of silicon wafers.

In addition to arc lamps, Vortek makes customized products such as liquid-cooled optical assemblies for research complete with supporting electronic modules; it has also developed sophisticated fabrication techniques for liquid-cooled tungsten electrodes. In-house computer facilities and proprietary software are used to engineer high-power optical systems.

In March 1993, Vortek held a demonstration for various emergency response groups and police and fire officials at the coast guard's hovercraft base on Sea Island near the Vancouver airport. Executives said later that if they could change the whole aspect of search and rescue — allowing crews to search all night using Vortek lamps — it would be "a major market opportunity." In May, Vortek was hoping to obtain between $500,000 and $1 million from the government to develop a prototype lamp that could be fitted on a hovercraft, and maybe even on helicopters. Reaction to their demonstration was favorable, so they began to prepare a formal presentation for defence and coast guard authorities in Ottawa in late May 1993.

The area that was lit up during the test measured 1.6 km wide. The demo lamp, one of Vortek's older models, was actually built for indoor applications. But even this model is 50 to 100 times brighter than the Zenon search light currently in use by hovercraft and rescue helicopters, according to Albach. The Zenon lights cost $30,000 each but cast only a beam of light; Vortek lamps deliver a wide-area light that costs $100,000 to $150,000. "Costs increase by a factor of four or five, but you get at least a 50 or 100 times increase in power," Albach said at the time.

Setbacks and Mistakes

Vortek's executives haven't had to learn much about managing change, but they have learned the hard way about creating markets, about changing products to suit markets, and about planning for the future by anticipating what more your product can do.

David Camm's advice, based on his experience in the school of hard knocks, is this: Check sooner to make sure there is actually a market for your product. In retrospect, he says, they probably should have done a little more research. "If we had, we would either have identified a market, or come to the conclusion that there was no market — in which case we would either have changed what we were doing, or folded it up."

The most important thing in any high-tech organization, says Camm, is not to get too tied up in the day-to-day operations of the company. "What you really have to do is keep your strategic goal in mind. Why are you in business? What are you trying to do? Where are you trying to get? If you know where you're trying to go, and everybody agrees on it and works toward it, then you're going to get somewhere."

Not losing sight of the larger picture is also important, says Camm.

> Business people have a saying: "Scientists should be on tap, not on top."

"If you don't have a long-term goal, then you'll spend all your time solving interesting little problems and wandering around in small circles and never really going anywhere ... Some people think globally, and others don't, and there's everything in between. In any organization, you need both. Scientists tend not to be global thinkers. They like to solve the problem in front of them. You've really got to keep your strategic goals in mind."

Bring in experts, if necessary, says Camm; and don't lose sight of your overall market potential. "Keep a global perspective."

He can't stress enough the importance of having a market that's sharply identified and clearly interested. "In order to get strongly increasing sales, what we'd have to do is identify a significant market which we'd then begin to penetrate," says Camm. "And unfortunately we have never identified one." He says that is the goal they're working towards now; and he has this to add: "If a market had existed for our original idea, we'd be outrageously successful now."

Risks and Choices

There have been years when the company experienced no sales at all, and hand-to-mouth phases when Camm and Albach thought the whole venture might go under. "It all just depends on when sales occur," says Camm. "We've gone from a year with no sales at all to one with the best we've ever had, to another with very low sales. It's unpredictable."

He does say the last two years have been fairly strong. "If all the people who say they'll be giving us contracts actually come through, this year [1994] will be better than any we've ever had."

Because of their product's unusually steep profit margins, one contract can still make or break the company, says Camm. For instance, in 1991 it looked as though Vortek might have to fold, but that didn't happen. What saved them? "We got a contract," says Camm, simply. "The first of those 1.5 megawatt lamps."

The Outlook for the Future

Vortek sees potential to expand its market in two areas: heat applications and search-and-rescue missions. At least one factor is working in their favour: the firm faces no competition within its market, "probably because it's such a small market," says Camm.

Their biggest market right now, says Camm, is high-temperature thermal testing, as in the case of the US Air Force and the American aerospace program: testing hypersonic aircraft. "The shuttle is to be replaced by an actual plane that takes off from an airport and flies up and starts off as a jet, and when it gets to the top of the atmosphere there's a little, tiny rocket that just flies up to the space station. This plane takes off and goes at about 22 times the speed of sound. They're scheduling it for about the year 2000," says Camm, "but right now they're testing it. In order to simulate the re-entry environment [of the jet coming back towards earth], they have to build high-powered irradiation units. That's where we come in."

That market opened up around 1986, and is still growing, says Camm. "We've completed the first stage of the contract; the next stage involves supplying four 1.5 megawatt lamps. Stadium lights you're used to seeing are probably 1,000 to 2,000 watts. So we're talking about the equivalent of 1,500 of those lamps, contained in something that's about a foot long. These would be used for testing engine components."

Their most crucial customer now is the US Air Force. About half of Vortek's market is in the United States, and the other half is evenly divided between Europe and the Orient.

The future will probably lie, says Camm, in surface treating metals or in thermally treating thin material. "Because we can deliver a large amount of radiant heat, we can actually melt steel."

> The market can be far removed from the original intention.

The Competition

Manufacturers of laser products will compete with Vortek to a certain extent, but Vortek can also do things no laser can touch, so the competition is actually fairly scarce. Vortek has a unique product, and to that extent, that it has created a niche.

Vortek has no competitors in its existing markets. If it starts to compete with thermal metal hardening, it will have entered laser territory, and will then begin to compete with some fairly established companies. But Vortek hasn't yet decided how to handle this competition, since it's still a non-issue.

"We are pursuing a market where we'll be using our lamp to treat metal surfaces, similar to a laser," says Camm. "If that takes off, then we'll expect to see competitors, because we're going to start to make a lot of money."

Management Strategy

In the beginning, Vortek was definitely a technology transfer company — high tech looking for a market. That remains true to a certain degree, but the company has also evolved to a point where it is actively searching out new markets to which it can tailor its technology.

The company has yet to experience a rapid growth phase. It's had some ups and downs, but over the last 20 years or so, sales have remained virtually unchanged from year to year, with sporadic highs and lows. "When we have a good year, we have a really good year," says Camm, "because our product is so expensive. In a bad year, we might not make any sales at all."

Vortek only recently put together a "business plan" of sorts. "We did the weekend retreat thing," says Camm. "We hammered out a company code of ethics and a mission statement." The two are handwritten in colourful ink and displayed in a plastic frame in Camm's office.

Camm admits the company's real need — its biggest challenge — is to identify new markets. About finding a market for the lamp as a source of light, he says: "Either there isn't one, or we haven't been doing it very well to date." They plan to put more emphasis on marketing in the near future. "There just may not be a market, in which case it doesn't matter how good you are," notes Camm. "Especially in our case, when it's technology push, instead of having the market dictate the product.

> Good technology must be combined with an understanding of markets and a serious business plan for the development of the technology.

"Doing a technology push is the wrong way around. You've got to find the market, then ask, 'What can we do to fill it?' And if you're working in an area you know something about, it's a lot easier. You make the products to suit the market."

They now manufacture products related to the lamp, though not really different. They also do a little consulting on the side in the reflector area, says Camm, but the lamp and its close cousins are really their only products, and have been for 20 years.

Vortek's greatest strength, says Camm, has always been its technology. "We have a very high profit margin, because there's no one else to do the job."

4

Quadra Logic Technologies Inc.: The Rough and Tumble of Focusing on a Direction

QLT is a biotechnology company that spent a lot of time experimenting with different research avenues and product lines before selecting one product, Photofrin, a new cancer treatment product on which to "bet the company."

In Quadra Logic Technologies's early days, its five founders — who started the company with the intention of raising and selling white laboratory rats — occupied a cramped office space on top of a Vancouver bakery, next to a primal scream therapist. Since then QLT, with 75 employees, has become a world leader in photodynamic therapy for the treatment of cancer, and its key product, Photofrin, is the first light-activated cancer treatment drug in the world to be approved for such use.

In those early days of primal screams and danishes, the company explored a diverse series of other products, from pregnancy and ovulation tests to generic pharmaceuticals. The company now maintains that its energies weren't scattered indiscriminately, but that there was instead a deliberate business plan to maintain a broad focus in order to feel out the different possibilities. Some seven or eight years elapsed between the original rat-raising plan and what eventually became QLT's current focus: photodynamic methods of treating cancer and other diseases, to be marketed worldwide within the next few years.

Now a mid-sized forerunner in the field of biomedical technology, QLT got its start when Julia Levy, Anthony Phillips, Jim Miller, John Brown and David Dolphin — four scientists at the University of

British Columbia and one business consultant — decided they wanted to set up their own business.

The original idea to sell laboratory rats sprang from their daily work in labs, for which they constantly needed rats. "As scientific researchers, they were all paying a phenomenal amount of money for the rats they used in experiments," says David Main, currently the company's director of investor relations and corporate communications. "These rats came from a very special island in the eastern United States, somewhere near Maine. There's nothing special about these rats — there's just some company on the east coast that's cornered the market on supplying laboratory rats.

"So they came up with the idea of setting up similar facilities on Quadra Island (just off of Vancouver Island, east of Campbell River) and selling rats to research institutions on the west coast."

What's the connection between rats and leading-edge biomedical technology to treat cancer? Levy, one of the five original founders and the only one still with the company, was an immunologist. Even while she was setting up the rat scheme, she remained involved in ongoing research at UBC on monoclonal antibodies (highly specialized chemical substances that target specific tumours and hormones). When it became apparent that the rat business was not as lively as expected, Levy and her colleagues decided to keep the company, but change its direction. They started branching out into diagnostics, generic pharmaceuticals and veterinary medicine.

"Tenacity, resilience and persistence are three adjectives that have to go in front of the word entrepreneur," says Main, who as a former practicing pharmacist owned his own business before coming to QLT. "What's never ceased to amaze me about entrepreneurs is the way they start off with a business venture, and if it's failing, they don't give up; they take a different tangent, so where you end up five years later is completely different from what you started out doing. It's because they're always looking for a win."

It now looks as though QLT might have one in Photofrin, the patented, light-activated drug for treatment of bladder, lung and esophageal cancer that has become the company's key product. Taken intravenously, it homes to cancerous tissue and is then activated by a laser to consume tumours without harming nearby healthy tissue. QLT received approval to market Photofrin from the Canadian Health Protection Branch (HPB) in April 1993, and predicted the product would be fully launched in Canada by late 1994. It has submitted similar applications across Europe, and finalized its application to the United States Food and Drug Administration (FDA) in early 1994. After more than a dozen years spent spinning off in different directions and eventually developing Photofrin, QLT looks set to take off.

History and Milestones

For the first five years — 1981 to 1986 — "not a lot happened," says Main. The five founders occupied a tiny office and worked only part-time at the new company, continuing to devote most of their energies to research at UBC. "They spent most of their time trying to decide exactly what they were going to do as a business and trying to find the money to stay afloat," says Main.

Around 1985, they decided to get serious. They spent more and more time on the company and concentrated on raising money. In the beginning, each of the five had put up $50,000 of their own money to launch the company; interested friends and family threw in a further $950,000. Private investors began to chip in. By 1986, having evolved from rats to pregnancy and ovulation tests to be marketed in China, the company decided to "become a full-fledged business" and go public, says Main. QLT began trading on the Vancouver Stock Exchange in 1986 for $2.50 a share; within a year, shares were worth $5, and the company had raised $9 million.

By about 1988, there were three main product areas: diagnostics (for pregnancy, ovulation and various forms of cancer); generic drugs; and veterinary pharmaceuticals. In the background lurked therapeutics for human disease, which would emerge as the most lucrative of all.

Before it became apparent that therapeutics was the place to pin their hopes, the company spent several years setting up strategic alliances and joint ventures with a slew of different international companies in attempts to further their existing projects. The first such partnership was with the Shanghai Institute of Planned Parenthood in China, which agreed to market a million pregnancy tests for QLT. The test consisted of a test-tube coated with monoclonal antibodies, and required users simply to add a few drops of urine and then watch for a colour change. The test took three minutes to perform, could detect pregnancy seven to twelve days after conception, and was 99 percent accurate. The ovulation test worked on the same principle and could predict ovulation 12 to 24 hours before the fact.

The technology was topnotch, but it soon became apparent that the market was already saturated. This arm of the company was eventually dropped, says Main, because QLT was too late getting into the market to be competitive, and sales weren't meeting expenses.

That same year, in an attempt to nurture the generic pharmaceuticals area, QLT considered a second joint venture with another Asian company, Guandong Enterprises Corp. They hoped that a joint venture could capture a 10 percent market share of the $1.5 billion market for pharmaceuticals in North America. The drugs — antibiotics and painrelievers, mostly — would be manufactured in Vancouver from raw materials imported from China. Jim Miller, president and CEO at the time, predicted competitors in the field would be hot on his heels in no time.

Unfortunately, this second prediction was more accurate than the optimistic first. The generic pharmaceuticals section of QLT was eventually dropped because of too much competition and low profit margins.

Turning from generic drugs to diagnostics, QLT announced in 1987 that it would link up with an Italian company, Industria Farmaceutica Cosmetica Italiana, to develop medical tests to diagnose a variety of conditions, including pregnancy and AIDS. Like its two predecessors, this venture was also doomed to be dropped after it became apparent that it wouldn't be highly profitable.

> Partnerships can be important but there is no guarantee that they will be successful, especially if the market is full of similar products.

"They decided the only way to make it big was to go for a new drug for the treatment of cancer, where there's lots of need, not a lot of competition, good patent protection, and good product definition," explains Main. By 1988, having reported only negligible sales since its inception, QLT decided to dispense with its other business areas entirely and focus on cancer treatment.

"Part of being a small, emerging company is having access to money," Main points out. "The only way you can access that kind of money is by having a good idea. Those other businesses were not ideas that people in the investment community were too excited about." All those former partnerships have now been divested, says Main — "sold off, discontinued, sold to other companies." That meant layoffs, too; before QLT freed itself of its wide variety of products, it was employing about 75 people. The number is about the same now, says Main, but it had to grow slowly back up after hitting a low of around 40 when the various businesses were sold off in 1988 and 1989.

Statistics indicate that taking a drug from the lab to the clinic requires some $230 million and 10 years, meaning that QLT has had to depend upon continual access to capital markets, showing off its successes and squeezing more and more money out of investors to keep on getting to the next step.

Late in 1987, QLT announced two major financings and an acquisition. The company bought 100 percent of Photomedica Inc., a subsidiary of the New Jersey-based medical industry giant Johnson and Johnson. At the time, Photomedica was already at the advanced stage of testing a patented cancer treatment drug called Photofrin II — the drug that has since become the key player in QLT's current success — in 40 North American medical clinics. Around the same time, American Cyanamid Co. of New Jersey, a major synthetics and chemicals company listed on the New York Stock Exchange, agreed to acquire up to a US$15 million equity interest in QLT over a three-year period (a total 17 percent interest). The deal required QLT to sign over worldwide marketing rights to its photodynamic therapy products.

> The regulatory process for pharmaceuticals is laborious and requires that firms have very "deep pockets" to survive.

By early 1988, the founders still owned about 38 percent of the company, which was now employing 57 people, half of them with PhDs. The company began trading on both the TSE and NASDAQ, and occupied itself with continued research and development. One year later, still primarily focused on developing and marketing Photofrin, QLT announced US patent protection for its second-generation family of light-activated benzoporphyrin derivatives (BPDs). BPD is a chemical cousin to Photofrin — different in composition, but intended for the same purpose. Photofrin was now in the advanced clinical testing stage, being tested for lung, esophageal and bladder cancer at 45 North American medical centres. The company announced a net loss for 1988 of $8.9 million, a leap from the $2.9 million loss in the previous year.

In the summer of 1989, QLT filed a new drug application with the Canadian HPB for Photofrin. By fall of that year, the company was still waiting for worldwide approval to market the drug. It was down to 50 employees and had about $10 million in the bank. Management ownership was down to about 22 percent of the company. Cyanamid owned a further 15 percent. In December 1989, QLT offered a further 2.4 million common shares to the public, half in Canada and half in the United States. This new stock raised the number of outstanding shares to 12.4 million and raised $26.9 million, which the company intended to use for developing, testing, producing and marketing Photofrin. By this time, the drug had been designated an "orphan drug" for the purposes of treating bladder cancer in the United States, meaning that the patented drug would enjoy seven years of exclusivity on the market as well as eligibility for tax credits and grants. Miller, still president and CEO, was estimating eventual annual sales of US$500 million following worldwide regulatory approval of Photofrin. Meanwhile, the company's balance sheet at the end of fiscal 1989 showed a loss of $8.6 million.

In early 1990, the company still had C$33 million in cash and a $10 million credit line with Cyanamid, enough to carry it through further research and development until marketing of Photofrin could begin. Investors were predicting profits by 1992. Losses by the end of 1990 were up slightly from the previous year, weighing in at $10.2 million.

During 1991 there were many shakeups at QLT. Jim Miller quit as CEO and president, and his departure was quickly followed by that of the VP of clinical development. By the end of the year, Miller had been replaced by William Foran, the former CEO of Cyanamid Canada Inc., a large subsidiary of American Cyanamid.

> At some point entrepreneurial firms have to shift to professional management to guide them through the next phase of development. Major investors usually insist on it to protect their investment.

Foran was a 40-year industry veteran who had joined Cyanamid in 1953, climbing from sales to management to presidency by 1977. He is credited with transforming Cyanamid Canada from a junior chemical producer to a diversified pharmaceutical firm. He was just what QLT had in mind — someone with the experience and know-how to take QLT through the upcoming regulatory hurdles and rapid growth phase. Miller remained as a director, holding on to about 4 percent of the company, which declared an $11.1 million loss that year.

"Jim decided to turn the company over to professional management," Main says now. "QLT needed somebody who had operating skills — not necessarily someone who was a mover and shaker, but who managed people really well and knew the industry enough to come in and take the reins."

Miller runs another start-up company now. "He's off doing what he does best, pulling the early stage concepts together, getting people excited about new technology, getting all the money together," says

Main. "That's a skill that some of the most seasoned managers don't have." Levy remains as senior vice-president and chief scientific officer. The other founders stayed on as directors until 1991, but have since left to pursue other new ventures, some backing Jim Miller's new project.

By 1992, QLT, with about $17 million in the bank, was spending money at the rate of around $1 million a month with no revenues and no cash flow. Said Foran at the time, "Clearly, we will run out of money if we don't go out and get some." He figured they needed to raise about $25 million to carry QLT through to commercialization, at which point revenues from Photofrin would begin to appear on the bottom line. The net loss for that year was just under $10 million.

Following HPB approval in early 1993, QLT elected to contract out manufacture of the drug to American Cyanamid, which would also handle marketing in Canada through its Canadian subsidiary. In early 1994, the price still hadn't been decided, but industry commentators were speculating a single treatment with Photofrin would be in the $1,500 range; some cancers would require multiple treatments. The drug was scheduled to be available in seven hospitals across Canada in late 1994, beginning with Toronto. These would be hospitals that already have the laser equipment necessary to activate the drug. Approval to market the drug to treat lung, bladder and esophageal cancers in eight European countries and Japan was still pending in 1994. Main estimates it will be another two years before they can start selling in the United States.

Currently, the percentage of revenues reinvested by QLT into research and development is "not a meaningful figure," says Main, because the company has no revenues to speak of. "We spend about $12 million a year on R&D," says Main. "That's everything."

Main says government funding, while still helpful, has ceased to be critical to the company's survival. "Relative to the dollar amounts we invest, what we get in government funding is very small. As we get bigger and move on, government funding becomes less and less critical."

However, he does underline the importance of government aid to fledgling start-up companies. "That can be one of their greatest sources of funding," says Main. "A lot of the research that government agencies will fund is not the kind of research that capital

---

While government support is important, even essential, in the beginning, a promising firm has to find other sources of financing (such as debt and equity) to get the amounts needed to grow. Private sector financial sources impose more business discipline than does public sector support.

markets will fund. It's too risky, too unsure. Capital markets spend their money on product research, moving a product through development."

Net losses for 1993 were $12.7 million, an increase of about $2.5 million over the previous year.

Setbacks and Mistakes

Around 1991, despite the fact that QLT had in the past divested itself of all interests outside of Photofrin, the company began to work with Baxter Healthcare Corp., a giant US firm, to develop a commercial blood treatment program that would give blood banks, hospitals and medical clinics a new measure of safety during transfusions. In retrospect, this looks like a desperate attempt to finally get the company off the ground; the company really wanted to have something to market by 1991, and it wasn't looking like Photofrin would be ready. However, it soon became obvious that this alliance would not result in profits for QLT. Still, the company took the disappointment in stride, chalking it up to yet another learning experience.

"We wanted to find out whether a certain approach to treating blood was a commercially viable process," says Main. "It wasn't. But it was a successful alliance in that it accomplished what we wanted, which was to know how viable the idea was.

"You should never look at partnerships like this as your lifeline," he says. "It's like a marriage; if it doesn't work out, you have to separate. Of course it would have been better for everybody if it had worked out, but just because it didn't, it wasn't a bad thing."

Main says in the beginning, the company had an intentional business plan to be diversified in four different areas; the hope was that each area would generate different but significant amounts of revenue and lead to a variety of opportunities. "There were people in the company who had expertise in those areas," he says. "It just happened that therapeutics for human disease was the one that grew the fastest and showed the most potential. We said, 'Hey, we're spreading our resources a little too thin trying to do all these things. Why don't we focus on one?'

"Each of the areas was probably doing moderately well, but they weren't the home run hits that entrepreneurs like," says Main. "So now we're fully focused on photodynamic therapy for the treatment of cancer."

Probably the biggest setback the company has experienced has involved winning approval from various national health departments to market their key product. QLT first filed an application with the Canadian HPB to market Photofrin in June 1989. There was no response until December 1991 — when the HPB came back with a request for more details. This interfered with the company's momentum, sending managers into another prolonged period of frustrated waiting while they tried to keep shareholders from growing too impatient, always promising profits would come soon.

In early 1992, Foran announced he expected to start commercializing Photofrin by late 1992 or early 1993. "In the next two to four years we are going to be marketing the product globally," he said at the time. As it worked out, Photofrin will begin to be available for purchase in Canada in late 1994. The Netherlands approved its use there in early 1994; approval is still pending in other potential markets.

Risks and Choices

The big risk for any biomedical company is the necessity of jumping through a seemingly never-ending series of regulatory hoops — or as Main puts it, "getting past the FDAs of the world." The company is happy that Canada's HPB was the first to approve Photofrin, but the rest of the world still stands between QLT and profitability. Canada represents less than 5 percent of QLT's potential market; the United States and Europe represent 40 percent each, and Japan the remaining 20 percent. Agencies like the HPB and the FDA consider every new drug that comes across their doorsteps to be guilty until proven innocent, because they shoulder the blame when something goes wrong, but get none of the glory when a new drug works miracles.

> Canada can be an initial showcase for technology-intensive products, but firms have to move rapidly into the international market place.

A second challenge, says Main, is that developing new treatments for human disease is a lengthy, time-consuming process. "You need a lot of money to do it," he says. "Keeping the interest of the investment community alive to keep funding you, so that you can continue to meet all your milestones, is a big challenge."

Another potential setback is the fact that QLT can sell only to hospitals that already own the sophisticated laser equipment necessary to make Photofrin work. In Canada, that figure is fairly insignificant.

A further concern is the ongoing reduction in health care funding in Canada. American Cyanamid plans to spend several million dollars to launch Photofrin in Canada, and up to $50 million to commercialize it worldwide. It remains to be seen whether those costs will be recouped as government health care dollars are cut further and further back.

The Outlook for the Future

Main says QLT has a two-pronged plan of attack for the future. Half of the plan involves timely follow-up of Photofrin with a second cancer-fighting drug still in the development phase: BPD. BPD is another form of photodynamic therapy. It uses lasers, fibre optics and light-activated drugs to treat cancer. So far, it has shown promise in treating not only cancer, but heart disease, psoriasis and various sexually transmitted diseases, and it seems effective in destroying viruses in the bloodstream. It works in essentially the same way as Photofrin — by attaching its molecules, like a little army of hitchhikers, to abnormal tissue, which is then zapped by a laser without harm to surrounding tissue.

"BPD is the new improved version," says Main. "Any of the attributes in Photofrin that weren't highly desirable are being improved upon, or eliminated, making BPD even better. BPD is still probably about four years away from the market, whereas Photofrin is on the verge of being in the market."

The plan is to get the jump on the competition. There are only a few companies in the world currently researching similar products — one in Switzerland, one in Japan and a third in California. "They're slightly behind, but they're close," says Main. "But that's where we feel we have the advantage. With Photofrin being first into the market you get first mover advantages, you get all the brand name recognition, reputation, all those things. Once Photofrin succeeds, all we have to do is follow it up with BPD."

The second part of the plan concerns ongoing research into related applications of Photofrin and BPD, especially in autoimmune disorders. Researchers at QLT are examining ways to eliminate cells in the bloodstream that cause autoimmune conditions such as arthritis

> Being "first to market" is a major driving force that shapes the business strategy of firms. Many Canadian firms have a "second to market" strategy, which is rationalized by saying that the "first-to-market" firm will make all the costly mistakes and pave the way. But then the rewards may be fewer.

and multiple sclerosis. The potential to broaden the possibilities for use of QLT's photodynamic treatments may lead to more strategic alliances in the future.

"Partnership with large multinational pharmaceutical companies is an important step in our growth," says Main. "There are multiple ways of using the technology we have, so we'll look into developing partnerships with other companies for the treatment of other diseases using the same kind of technology."

To that end, QLT signed a licensing and research agreement in August 1993 with Baylor Research Institute of Dallas, Texas, to further studies of viral inactivation and the treatment of blood-borne diseases. The agreement is intended to help QLT explore promising new applications for photodynamic therapy with BPD.

The Competition

QLT's biggest competition, says Main, will come not from companies directly competing with similar or advanced technology, but from the old technologies still in use, whose manufacturers claimed that market niche as their own years ago.

"We deal with two forms of competition," says Main. "One is existing products that doctors are using to treat the conditions we want to treat; and the other is people developing similar products. We have a really good lead, we have the world's most renowned porphyrin chemist as one of our vice-presidents, we have a world-renowned immunologist — we have a solid scientific core and our knowledge base in porphyrin is probably 10 times that of anyone else.

"That's not to say people can't catch up, but we don't want to squander that lead, and we're staying well ahead.

"Our biggest challenge will be convincing doctors — who are used to treating cancers a certain way — to adopt a new approach." He says the company has no worries about how its product will be received in the market because they already have so much insight into what doctors think about it. "It's been 'in use' for more than five years now in clinical trials," says Main. "So our customers are not unknown to us. The clinical trial process helps build relationships with potential customers."

Main says it's hard to predict what market share QLT will have once the worldwide market opens up. "It's difficult because there are no precedents," he says. "It's not another drug for the treatment of cancer — it's a whole new approach. So we'll start with low market penetration and build our way up over the years. Adoption of new technology is a slow process." Working in QLT's favour is Photofrin's relatively low cost — $1,500 per treatment, compared to $30,000 to $40,000 for existing treatments such as chemotherapy or surgery.

He will say this much: "If we can penetrate our market by 10 percent over the next three years, we'll be happy." They should be: the overall market is estimated to be worth up to $500 million worldwide.

Management Strategy

The company's greatest strength, says Main, is its people and their science. "The average IQ in this company is probably about 30 points higher than anywhere I've ever worked," he says. About half the staff have PhDs. "They're all really driven, hardworking, highly educated people. Because of that, there's an energy here, people really want to get things done."

Gearing up for the rapid growth set to happen the moment regulatory hurdles have been cleared has meant changes in management strategy that have also affected the nature of the business. The company has seen a shift from inexperienced entrepreneurial spirit to sophisticated professional management.

> The need to move from an entrepreneurial mode of operation to a more disciplined professional mode can be very difficult for some entrepreneurs.

"The whole board of directors has essentially changed," says Main, "going from a roomful of local entrepreneurs trying to make sure the company survived, to a board of mostly seasoned business professionals, a large percentage of whom are from the pharmaceutical industry, to give us stronger industry and market focus."

William Foran, CEO and president, will oversee the next few years of rapid progress by maintaining a highly qualified staff. "The first thing he did was make sure we had the right people in place to cover off all the areas of the company that need expertise, so that we wouldn't have to rely too heavily on our strategic partner [American Cyanamid]," says Main. "He wanted to have that same expertise here, so he could make sure they were doing the right things for us and make good planning decisions."

Keeping people with senior experience in marketing and operations, manufacturing and other key areas is of paramount importance now. "We still don't have the depth of a large manufacturing company, but we have that key expertise so that we can manage our business and make good planning decisions for the future, knowing the critical success factors in our industry," Main says.

Of course, as director of investor relations, Main is quick to point out that on the capital side of things, it's important to maintain positive relationships with the investment community. "We have to keep our customers happy, but we also have to keep our investors happy and informed," he says. "And we take that very seriously."

5

Xillix Technologies Corp.: Matching an Opportunity with Seasoned Management

Xillix is a start-up company that is commercializing leading-edge medical research. The original entrepreneurs recognized early on that to be successful professional management was needed and they brought in a seasoned CEO.

"In a high-tech environment," says David Sutcliffe, president and CEO of Xillix Technologies, "there are going to be dramatic ups and there are going to be dramatic downs. The idea, if you're really good at this, is not to have the ups so up and the downs so down."

Sutcliffe will have ample opportunity to try his hand at being good at this, because in the next few years, as its products are launched one after the other, Xillix will be a prime example of a company trying to walk the fine line between staying competitive, with leading-edge technology, and trying to appease its shareholders and investors by showcasing its professional management skills and profits.

Xillix, a TSE-listed medical technologies company, was first formed in 1988 to commercialize research being conducted at the British Columbia Cancer Association (BCCA), which still receives a 1.5 percent royalty on Xillix's gross sales. It was founded by Branko Pacic and Bruno Jaggi, two researchers employed by the BCCA, and Michael Routtenberg, an entrepreneur who funded the company partly by not drawing a salary at first. Xillix was an amalgamation of two sister companies originally called Microscan and Endoscan. The founders decided it would be easier to attract financing if everything was under the same umbrella, says David Sutcliffe, president and CEO. "Also, it wasn't ideal to have all the corporate infrastructure repeated twice and costing you twice. So the two were amalgamated into one, which was eventually named Xillix." The two companies were each founded in 1988, then amalgamated in 1990 and called Microscan Imaging and Instrumentation Incorporated. The name change to Xillix happened in 1991 after legal problems with previously established companies that had names similar to Microscan.

But the name game is a losing battle, says Sutcliffe, who is still slightly amused about all the red tape. "We got the name Xillix, had it approved as a registered trademark in the United States and Canada — and it turns out there's an American company based in California and trading on NASDAQ, called Xillinx," says Sutcliffe. "It's about a US$180 million per year company in electronics." The moral of the story, he says, is that "no matter how hard you try to have a unique name, there'll be some conflict." They chose the name Xillix in the first place, says Sutcliffe, specifically because it didn't mean anything.

In 1994 Xillix has three main products in various stages of development, production and marketing: MicroImager, Access and Life. "MicroImager" is a patented, computer-based digital camera used to study cells, and is most useful in fluorescent cell imaging — where illumination is low — for examining large areas for suspicious cells. It captures and produces high-quality fluorescence images, making it particularly useful in genetic research and screening.

"Access" is an automated cervical cell prescreening system designed for automated analysis of pap smears. These smears are currently examined by cytotechnologists, who view individual specimens through a compound microscope. These technicians are in short supply, command high salaries, need intensive training, and are subject to large margins of error because of the long periods of concentration and judgment required by their jobs. Access would replace cytotechnicians, speeding up the process, making it more economical in the long run and reducing the rate of false results. In the United States, some 70 million of these tests were analyzed in 1991. About 60,000 women in the United States develop cervical cancer every year.

"Life" is a lung imaging fluorescence endoscope used to detect lung cancer in its early stages. It is an imaging device that measures the difference between the intensity of autofluorescence in normal tissues and cancerous or pre-cancerous tissues. In North America, about 180,000 new cases of lung cancer are diagnosed each year.

So far, MicroImager is the only Xillix product that has been launched commercially. Access is still in clinical trials, and Life has been partially launched — it's still in clinical trials in North America, but is selling commercially abroad and, say managers, doing quite well so far. Xillix signed up 10 dealers and distributors in Asia during 1993.

Customers for Life, when it is fully launched, will be thoracic surgeons and pulmonologists; Access will sell to pathologists and clinical research labs.

Sutcliffe, who came to Xillix in March 1992, has a background that includes both software engineering and managing product development. He also brought some marketing savvy, gleaned from his time at Motorola as VP marketing, to the company. His career path took him from computer science at the University of British Columbia to a Vancouver high-tech company called Sydney Development, which he joined when it had about a dozen employees and less than $1 million in sales. He stayed for seven years, saw the company grow to 300 employees and $25 million in sales, and then watched it plummet almost all the way back down again, until it eventually went out of business. From there, he went to Motorola's Mobile Data Division (formerly Motorola Data International Inc.), where he stayed for four years until he was wooed by Xillix.

> Unlike many start-up situations, where the founders hold on in the belief that their technology will sell itself, Xillix saw the need for professional management very early in the firm's life.

"I came to Xillix for the opportunity to build a world-class business right here in Vancouver, or more generally, in Canada," says Sutcliffe. "I don't want to have to move out of Canada to pursue my career. Xillix found me, and it was an attractive offer — to come and be president of an emerging company that had all the ingredients to become world class. We've still got to prove we can be successful and profitable."

Sutcliffe insists that it's more than Vancouver's vast quality-of-life appeal that keeps him here; it's more a personal mission to prove that Canada can compete on a global level. "People in this industry are annoyed by the media portrayal of Canadian high tech as being not quite critical mass," says Sutcliffe. "Companies that do become successful often end up getting bought by multinationals," which means if you want to be on a fast-track career path, you are certain to end up living outside of Canada — and that shouldn't be necessary, he says.

"I get annoyed when I hear people say that, well, the only way to be successful in a Canadian technology company is to sell the company to an American company or a European or Japanese company. I don't believe that. What I'm doing here is demonstrating that that doesn't have to be true."

History and Milestones

The investment that fed Xillix from 1988 to 1992 came from two primary sources: private investors ("angels," says Sutcliffe — "well-to-do private individuals who elected to invest in the company"); and funding from provincial and federal government programs: Industry, Science and Technology Canada, the Western Economic Diversification Fund (WDF — a program established by the federal government in 1987 to diversify the Western Canadian economy by encouraging new products, markets, technology and improved productivity), and the Ministry of Economic Development (now the Ministry of Employment and Investment). Sutcliffe says the money came about equally from both sources.

> Multiple sources of significant funding can be especially important to a start-up firm in the medical field that faces costly clinical trials so that their products can meet regulatory requirements.

"Then, around 1992 we did a private placement," says Sutcliffe, "using the services of two underwriters and raising $5 million, the largest financing at the time. Then in October of that year we made an initial public offering on the Toronto Stock Exchange and raised another $13 million." Shares began trading at $4 each in September 1992.

That's still where most of their financing comes from now, along with revenues from sales. "We're still getting some government support, but almost all of our funding now comes from public and private shareholding investors," says Sutcliffe.

All three original founders are still with the company, in varying capacities. Branko Pacic, who has a PhD in biophysics from McMaster University as well as a degree in engineering physics from the University of Ljubljana, Slovenia, is on the board of directors and is chief scientist. He still teaches at UBC in both the medicine and physics faculties. Bruno Jaggi had been spending most of his time recently with the BCCA, but at the beginning of October 1993 he took an extended leave of absence from the agency and joined Xillix again. "He's now our VP and chief engineer, responsible for all product development and manufacturing," says Sutcliffe. "So he's even more involved today than he was before."

Michael Routtenberg, an entrepreneur before he started Xillix, was originally the president, and became a vice-president and director when Sutcliffe came on board. He's still a director, but has stepped down from an active role in the business because he's off to do his next start-up, says Sutcliffe.

"He's out working on a business plan, not drawing a salary, trying to get another business off the ground. He spent five years with Xillix — he got it going from where there was no office, no company, no salary, all the way through to where it's a public company with dozens of employees."

Bringing in managers with previous track records has "strengthened, reinforced and enhanced" the company, say executives. When the company finally reached the critical stage where marketing smarts and skilled management were becoming increasingly crucial, the three original founders brought in experienced managers in the hopes of pushing the company to further levels of accomplishment. The idea was to enable Xillix to move forward to the next step with the "usual bureaucratic departments" in place, say managers.

The first such executive Xillix imported was Barclay Isherwood, who became a director and board chairman in early 1992. Xillix is banking on Isherwood's previous track record at Mobile Data International, where he was president and CEO for almost a decade. Prior to that, he was VP marketing and eventually vice-president and chief operating officer at MacDonald Dettwiler & Associates Ltd., a highly successful Vancouver high-tech firm. He has a master's degree in engineering from UBC.

Xillix's second managerial acquisition, two months later, was David Sutcliffe. The company is relying on these two high-tech management veterans to lead the company into profitability and to inspire confidence in investors and shareholders.

The company's first customers were cancer research labs, mostly in the United States, for the MicroImager. To no one's surprise, according to Xillix, that first product was received enthusiastically due to a "pent-up" demand for its microscopy applications.

The fact that those early customers are so satisfied, says Sutcliffe, bodes well for the products that will follow. "Early customers are absolutely crucial — not only for the obvious reason that they help you get your product right, but because those customers become your best salespeople," says Sutcliffe. "Better than any salespeople you could ever hire. A customer who bought your product and uses it and believes in it will champion that product far more credibly than any paid salesperson to other customers. Those early customers are critical for getting reference sales, getting your sales going in general.

It's crucial to have a first product that is top quality because, as Sutcliffe points out, "you can't hire a customer." Other customers, he says, will believe your first customers.

> A successful first product launch establishes a firm's reputation.

The worldwide potential market for Xillix's three products could total in excess of $1 billion if estimates prove accurate. The breakdown would be $400 million for Life, from $600 million to as much as $1.4 billion for Access, and $20 million for the MicroImager.

Only a fraction of that market is in Canada — around 1 percent. The rest is split almost evenly between Europe, the United States and Asia.

> Knowledge-based firms have to get into international markets rapidly.

Xillix is currently spending some $3 million a year on research and development, which involves about 60 percent of its staff, says Sutcliffe. "Continued R&D is absolutely critical to us. We're not a one-product company, nor are we going to be able to launch one version of our product and leave it the same and expect to be successful. So continued R&D, both to enhance and improve our existing products as well as to bring new products to market, is really critical."

Sutcliffe says he fully expects the current expenditure level to continue, although he points out that the percentage of Xillix's workforce involved in research and development will decrease relatively over time as expenses in sales, marketing and distribution increase.

The $3 million being spent on research and development right now surpasses Xillix's revenues, which are about $2.1 million, says Sutcliffe. "But I would certainly expect us to be at the 10 to 12 percent end of the spectrum in the foreseeable future."

Setbacks and Mistakes

"People talk about the S curve, the chaos theory," says Sutcliffe. "There are all kinds of theories for managing high-tech outfits. My own personal theory is the roller coaster." The key to not being thrown from your seat on the roller coaster, says Sutcliffe, is to "have the ups and downs level out, and have a trend line that keeps going up."

That would be nice in an ideal world, but Sutcliffe acknowledges that such a goal is not always attainable. "The truth is there are going to be ups and downs in the early stages of any business with these kinds of risks," he says. "You try not to get the rate of growth so high that you can't manage it, or inevitably something goes wrong and you get a big down."

He says he has seen companies forced to lay off vast numbers of people after a long period of rapid upswings followed by equally swift drops. "I've certainly had to lay off dozens — certainly laid off more than 100 people, and it's a traumatic experience for everybody involved," he says, referring to his time at Sydney Development. "We're strongly attempting to avoid that here."

Xillix had some close calls in its early years, he says. "There were a couple of points when the company ran right down to its last cash and had to raise money in order to be able to meet its payroll — not that long ago, either," says Sutcliffe. "A year ago when I joined Xillix, there was $150,000 in the bank and 42 people on the payroll. It doesn't take very long to go through $150,000 when you're meeting payroll for 42 people. And you can get into very vulnerable positions if you let your cash go down too low."

Risks and Choices

One recent decision managers now agree was for the best — even though it has taken up vast amounts of time and energy — was the choice Xillix made in 1992 to go public. The company opted for that route in order to finance the next stage of growth, say executives. The company now admits going public has taken up a lot of energy that might have been directed elsewhere, but they maintain it was worth it: the initial public offering raised $13.6 million.

> Getting the necessary working capital to create a stable environment for the firm's development is always a major challenge.

"We have enough cash now to allow us to focus on business more," they say. "We have a window of two years to invest in trials, marketing and manufacturing, as opposed to bootstrapping in order to build a business."

The company faces several challenges in the years to come. One, say managers, is missionary marketing; another is getting regulatory approval in a variety of countries, which could suspend production and marketing for years. A third will be convincing the medical establishment to switch to Xillix's technologies. Although the company has succeeded in carving itself out a relatively significant niche in the market, it still has a long way to go. That's because "the medical community is very entrenched," says one manager. "The lead times for acceptance of a product in the medical market is much longer than for most other consumer products."

As with any company focusing on medical or biomedical technology, a big risk factor is rising health care costs. Concerns with the skyrocketing price tag on health care could cause governments and insurance companies to grow increasingly conservative in their health care spending, which could have a large impact on the potential size of Xillix's market as well as its profitability.

One of Xillix's tactics for avoiding disaster is not making financial projections. "We've been extremely careful not to make multi-year, high-growth-rate projections — not because we don't believe we'll attain that kind of growth, but because we don't want to be guessing and then significantly disappointing our investors by missing them in whichever direction," says Sutcliffe. Total revenues were $2.1 million in fiscal 1993, and were projected to reach about $4 million in fiscal 1994.

The Outlook for the Future

Xillix's greatest strength, say company executives, will continue to be breakthrough, leading-edge technology combined with seasoned management.

Xillix's focus for the next few years is going to be aggressive commercialization of its existing product portfolio. "We aren't yet a profitable company," says Sutcliffe. "We're growing up a rapid growth curve, but we have to start producing profits. We have to complete remaining development and clinical trial work and build our distribution channels."

Life is on the market today in some international markets, but it's still waiting for approval from the United States Food and Drug Administration (FDA). "It's a lengthy process and we don't know when we'll get approval," says Sutcliffe. "But we can market it in Canada. And it's on the market in a number of countries now."

Access is on the market internationally for research applications only; clinical applications still need some additional work, and that will happen internationally well before it happens in the United States.

"It's frustrating," says Sutcliffe. "It's a major barrier to innovation. The United States represents about 40 percent of the world market for sophisticated medical equipment, and we can't access that market effectively until we have FDA approval."

The future also holds the possibility of strategic alliances with other similar businesses, says Sutcliffe. "It's inevitable that we'll develop closer relationships with some similar companies, because our product has complementary positioning with some other technology and products that are out there," he says. "So we'll need to make friends with some big companies to make sure that our product line is seen as complementing or enhancing their product line — so that we've got a good market environment to go sell in."

"We'll also continue to be quite proudly Canadian," he adds, "so these relationships will likely be just that — relationships."

There are no plans yet to develop any completely new products, but the company does envision many other applications for its existing ones. Access, for instance, is a software application that could be used to screen for other types of cancer, such as breast cancer.

The Competition

The MicroImager's main competitors are companies — such as Kodak, Photometrics and Hamamatsu — participating in a market that is similar to, but broader than, Xillix's. Xillix focuses on high-resolution visual imaging, a much more narrow niche than what most of the competition is involved in; it specifically focuses on microscopy, which is only one component of a much broader market.

The level of competition is different for each product. With Life there is no existing system for early detection of lung cancer that is remotely comparable, even in development stages. Xillix is aware of several companies that are currently investigating similar technology, but because of the need for regulatory approval, Xillix is still giant steps ahead of any such competitors.

Access is a different story. Xillix estimates that there are at least half a dozen other companies aiming to develop technology that would automate pap testing, and some of those may already have filed applications with the FDA. Xillix hopes it will edge competitors out of the game with its reasonable cost, high accuracy rates and unconventional approach (it classifies cells by focusing on the texture of nuclei).

For the company as a whole, Sutcliffe has a clearly defined idea of what will make Xillix competitive and enable it to stay several paces ahead of followers. "Putting all the critical pieces together — technology that's competitive, a market with a burning need for your product, the right people, access to investment capital — is what makes a company world class," he says. "When I joined Xillix, it had each of these things. I went down the checklist in my mind."

The difference between companies that will go on to be world class and those that won't, according to Sutcliffe, is that world class organizations combine all the ingredients of success.

"There's lots of investment money out there to be had," he says. "Investors say there just aren't enough good companies to invest in."

Management Strategy

Sutcliffe says by his estimations, Xillix is right in the middle of its first exciting rapid growth phase now. "Particularly as it pertains to revenues and customers," he says. "We've already spent those long, slow years getting going, getting all the investment and technology and products, and understanding the market and raising capital. In the last two years we've gone from no revenues to several million dollars in revenues. It's an exciting time and it's a pretty rapid way to grow. We're expecting that rate of growth to continue and maybe even accelerate for the next two or three years."

> A company that attracts investment is one that can combine technology, market and management. Too often, at least one of the ingredients is missing.

In Sutcliffe's opinion, there are three key things that will be important to weathering this growth period successfully. The first is people with impressive management experience. "We want to hire people with experience in both high growth rate businesses and in businesses that have been the size we want to be, as opposed to the size where we've been," he says. "That's a key thing. If you're going up that S curve you need people who have experience at both ends of it, ideally. That's very hard to find in a person, in practice. But I would stress that far above any other factor."

The second ingredient is management discipline. It's important, says Sutcliffe, for everyone on the team to understand how you're going to manage the growth and what the company's goals are. "It has to be a consciously explained process," he says.

And finally, capital is essential. "Probably the easiest thing to do when you're in a high rate of growth is not to pay attention to the company's requirements for cash," says Sutcliffe. "In businesses with high rates of growth, the faster you grow and the more successful you are, the more cash you need, not less — most of the time, anyway, in early stage technology companies."

He warns that it's dangerous to get lulled into being successful in a number of areas of the business and consequently forget that those successes will create even higher cash requirements. "So paying attention a year or two years out — rather than one payroll or one month out — to the cash flow for the business is a key thing."

And cash, in Xillix's case, requires stakeholders. "When you first establish a business like this," says Sutcliffe, "you're going to have a tremendous range of stakeholders, more than you ever envision — not just the obvious ones like employees and investors, but suppliers: your suppliers can make you or break you in the beginning. Recognize that right from the start: Who are those people and how are you going to create value for them?"

Set realistic expectations and then meet them, says Sutcliffe. "The invention of the spreadsheet has been referred to as one of the biggest boons in corporate financing, because anybody with a computer and a spreadsheet can make a hockey stick revenue projection model that says their company's going to have $100 million in sales in no time," says Sutcliffe. "It's too easy to sit down and do that. You have to set realistic expectations so you have a chance of meeting them and not disappointing your stakeholders.

A professional manager, unlike many entrepreneurs, sees the business plan for the firm as fundamental to success.

"And then you have to meet them, so that you gain credibility for setting expectations, getting people on side for them, and meeting them."

A business plan is an absolute, says Sutcliffe, who is quite adamant about it. "If you don't have a business plan you're not a business person," he says. "You're not conducting a business. If you don't have a business plan at the start-up stage, then you're not starting a business, you're starting a hobby."

One big pitfall to steer clear of, says Sutcliffe, is what he calls convenience relationships. "It's a big mistake to get involved in those kinds of relationships early on," he says. "They get in the way of the right ones later, and then you're really disadvantaged."

Entrepreneurs, says Sutcliffe, are constantly being told they have to have a strategic partner or a distribution partner to enter world markets or break in to the American market. "So the early-stage entrepreneur, not absolutely believing that notion but having it repeated to him by many people, goes out and finds the first people who warm up to his story, and appoints them as worldwide exclusive dealer or distributor, or gives them rights to the United States market," says Sutcliffe. "And often these deals are made out of convenience, just so the entrepreneur can say he has a distribution channel — instead of actually having a channel that works."

This happens quite frequently, says Sutcliffe. "The channel doesn't work, to no one's surprise, but you may have needed that channel to get the story right, so that a financier would come in and keep your business alive. The problem is you're robbing Peter to pay Paul. You've satisfied the need to check off that distribution channel box on your business plan and in your story, but of course the channel doesn't work, sales don't come out, and everybody thinks your product must be a dud, because if it was so great and you've got a channel, why isn't it selling? That happens on all levels.

"People pick their investors by that same method," continues Sutcliffe. "They pick their banker that way, their financial advisors, their lawyers, all kinds of things. If you want to be successful you really need to think about what the business needs, and go find the right partnerships and relationships. Selectivity is the key word here. Be really choosy, and stick to your guns. You've got to ask yourself: What's your vision for the business? What are the implications of that?"

6

Techware Systems: Techies Doing Their Own Thing

Techware is a highly specialized, technology-driven firm that is still managed by the founders, who believe that maintaining an entrepreneurial culture is key to success.

Richard McMahon is a westerner at heart, but he still remembers the summer he spent in Ottawa in 1979 as a turning point in his career. He had a summer job at Bell Northern Research (BNR). It set the direction he would take with his master's thesis later that year — and that project has turned into the backbone of the $4 million company he runs now.

Founded in 1983 by McMahon, Techware Systems is a private company that provides automation equipment and expertise to the vacuum equipment and thin film markets throughout North America and Europe. The company now employs 35 people in its new office space in Riverside Park (an industrial park in Richmond, BC), and boasts a growing number of sales representatives in Europe, Asia and the United States.

Techware was born when McMahon, a process engineer, decided to commercialize the inventions that came out of his master's thesis in engineering physics at the University of British Columbia. The thesis focused on the automation of processes related to microchip fabrication. He envisioned the development of a flexible, high-quality automation system to improve the quality of thin film manufacturing processes by achieving better process control. Straight out of the master's program, McMahon founded Techware with Michael Hanssman, an electrical engineer who designed the initial hardware product and who, as principal engineer, is still responsible for ongoing engineering developments at Techware.

McMahon says he first conceived the idea for this product while he was a summer student at BNR in Ottawa. "I was running a machine all summer, tweaking the knobs and reading the dials and going back and tweaking more knobs, trying to make a certain product," he says, "and somewhere through the summer it dawned on me that maybe a computer could do this better than I could." So he went back to UBC to start a master's degree in engineering physics, where his thesis project was building a computer to control "sputtering." Sputtering is one of several popular processes used in the microelectronics industry for depositing a microscopically thin layer of one material onto another material.

> Ideas for a technology venture can come from anywhere, even a summer job.

"Often these materials are fairly mundane," says McMahon, "for instance, aluminum — but they can also be fairly exotic materials, like titanium nitrate or other compounds. Generally the idea is you're depositing it atom by atom, building a thin layer, onto something else. It's like building silicon chips — you start with raw silicon and deposit a layer of an oxide, then etch that back and deposit a metal. Or you take a roll of mylar plastic and deposit a magnetic coating on it that can be magnetically charged. CDs, for instance, are a platter of some kind of plastic with layers on it that will react with the light and store the light information."

Similarly, thin films are used in architectural glass, such as office buildings with gold-plated windows. The gold is layered so thinly — about 10 atoms or so — that the light still filters through it. "You can't just paint it on," says McMahon. "So you use one of these vacuum-based, thin-film processes." Sputtering is one such process.

Techware's key customers now are giants like IBM, Motorola, Intel, AT&T, Honeywell, Digital Equipment and Xerox. The company still specializes in real-time scientific instruments and industrial process controllers. McMahon is projecting 1994 sales of $4.2 million.

History and Milestones

When McMahon graduated from UBC, he was hired by the university as a research associate to continue the research his thesis had begun. In the same year, he co-founded Techware by hiring Michael Hanssman, an engineer he knew from grad school. Hanssman worked alone for much of that year, while McMahon continued his research at UBC. Late in 1983, McMahon finally joined him.

Shortly after that, the two received a BC Science Council grant worth $80,000, enabling them to hire a third employee. Their efforts rapidly resulted in their first contract for more than just one single computer at a time, says McMahon. "The customer was a capital equipment manufacturer looking for a third-party control system. So we signed them up, and from that we were able to get enough revenue stream to hire additional people."

McMahon isn't surprised at how easily they were able to make a go out of commercializing his thesis project. "We were very entrepreneurial in the sense that the professor I worked for as a grad student [Dr. Robert Parsons, who is still with the university] was always thinking about commercialization," says McMahon. "He created an environment of entrepreneurial excitement in his lab. There was more than one company started out of that lab." Techware was Hanssman's first job out of university.

McMahon says that along with his experience that summer at BNR, his master's thesis idea originated in part from his interest in studying the possible applications of solar energy. "I was originally really interested in solar energy, and I thought these thin films could be used in solar energy," he says. "When I started my master's degree, really the project I was working on was automating equipment, but there were five other students in the lab, and the whole project was to make a coating transparent to light but reflective to infrared heat — the idea being that the sunlight could come in through a window coated with this, and warm up the couch, the floor, the room; then it would become infrared radiation and radiate back out. That infrared radiation, when it hit the window, would see it as a mirror because it would be reflective to that wavelength and therefore would bounce back in: transparent insulation for windows.

> An energetic champion can provide a catalytic influence. Professor Frederick Terman of Stanford University, one of the godfathers of Silicon Valley, advanced $538 to students William Hewlett and David Packard to pursue the commercialization of their research.

"That was the real project I was interested in; building a controller was something I thought would be useful in making those kinds of films."

As it turned out, there was more commercial interest in the computer control system for that project than there was in the results they were getting in heat mirrors themselves. McMahon and his colleagues noticed this as soon as they started publishing results. "There was academic interest," says McMahon, "but the interest said, 'Hey, where can I get a system like this, because this would be useful for my work, too.' So we got enough interest, and essentially I said, 'Let me incorporate a company and then tell you where you can get something like this,'" McMahon jokes. "That's the short story. We actually went out and raised some National Research Council money for transfer of technology from universities into commercial use." That was when McMahon was hired as a research associate at UBC, trying — as he says — "to turn this haywire laboratory computer system into something that would be more appealing to the real world." When they developed it into a more commercial form, they started getting some real customer interest, and they ran with it.

"We had no formal studies," McMahon says, "but we had been talking to a number of people in the scientific industry, talking to people at trade shows, that sort of thing. I had a contact at Stanford University, one at an IBM research centre, another at 3M, who all said yes, we're interested in buying one of these if you go with it. We formed the company, and Stanford University became our first customer."

Their first product — PAL-68000, a computer system that automates the process of microchip fabrication — was well received initially. But Stanford found that although the product did an excellent job of meeting its needs at the time, it wasn't flexible enough to be used far into the future. "Their ideas changed and they needed new things," says McMahon. "In other words, the control algorithms were hard-coded right into the software, making it very difficult for them to change their minds midstream and try out something new." Recognizing the need for flexibility, McMahon and Hanssman went to the BC Science Council with a proposal to develop what's called a fourth generation language — the idea being, as McMahon puts it, "Let's have a core platform that is identical for everybody, but the differences in research methodologies or in control algorithms or differences from exactly what one person wants to do to what the next person wants to do could be entered from a data-entry level, as you do on a spreadsheet. The controller would, therefore, be more flexible; the end user could get in and modify their algorithms without having to become a C programming guru."

That was where the company stood after two years. The starting capital of $50,000 that launched the company from that point on came from that first sale to Stanford University. "We were really fortunate," says McMahon. "They recognized what we had as being unique and were willing to pay for it." The university, according to McMahon, had some $50,000 left over at the end of their fiscal year, and they had to either spend it or lose it. "So they spent it on our product, and gave us a 90 percent prepayment," says McMahon. "For the first unit we built, we got a 40-some-thousand dollar cheque in the mail. That's probably the luckiest and most key financing we ever had."

That's because it produced a snowball effect — once they had a customer in line, they were able to get another customer, and then another, says McMahon, "and then we got the Science Council to treat us seriously." After that, IRAP chipped in, and Techware started to receive other, smaller grants from the National Research Council and the Defence Industry Production Program — a $300 million program in which the Canadian government matches a company's investment in research and development dollar for dollar — along with another BC Science Council grant and one from the Western Economic Diversification Fund. "There's been a lot of good government support," says McMahon.

Still, he says the government support they've received has ended up being only a small fraction of their overall revenue scheme. And last but not least, he says, "There's ourselves — family, friends, dogs, cousins, anybody we could find who had money they wanted to put into a retirement nest egg of some sort and help us out."

During the next eight years, the company grew in number to about 30 employees, dropped down to 14 during the late-80s recession, and has now climbed back up to 35. Techware still manufactures essentially the same product — a computer system used in the thin-film industry — although continued research and development has seen it improve and become more flexible and widely applicable over the years. In 1989, Techware moved from its tiny quarters in downtown Vancouver to a modern, spacious locale in an industrial park in Richmond, and McMahon is considering the possibility of another move in the near future, due to continued expansion.

Sales didn't start to reach significant figures until about 1987, when they clocked in at around half a million dollars. They grew quickly though, reaching $1.2 million by 1988 and doubling that figure in 1989. They took a slight dip during the recession and resurfaced around $3.4 million in 1993.

Techware now calls itself a leading supplier of control systems to the semi-conductor and high-vacuum industries. It focuses on delivering complete integrated solutions, accepting full responsibility to address a customer's control requirements, including the engineering expertise, ability to install, service and support the product, as well as the hardware and software necessary to provide a high quality solution.

"We are definitely past start up, probably somewhere in the middle of rapid growth," says McMahon. "I like to think of it as a calibration — one level, in technology, is initial sales, when you're really out on the leading edge and you're able to get a lot of money for your product, but only from a few select customers, really cream of the crop customers that want truly leading-edge stuff that's high-risk, high cost. It may be just exactly what they need. That's where we used to be.

> Finding a product niche is important as long as there is a sustained effort to ensure that the product continues to be improved.

"We've essentially fallen behind a little technologically, we're not really leading edge any more — now we're what you might call advanced technology, meaning we're still somewhat out front, but in a less risky business sense. We have a proven track record — we've sold 500 units around the world — and people can really bet their businesses on us. But on the other hand we're not radical, we're not way out there with something that's totally wild. So that puts us somewhere in the high growth area." He says the challenge will be to make sure they stay there, which means continued reinvestment in technology. "If we don't do enough R&D, the curve will keep going while we stay still."

As for research and development, says McMahon, the company currently reinvests about 15 percent of its revenues, and he doesn't expect that to change any time soon. "We're just trying to identify what our core competencies are," he says, "and what they'll have to be, in a five-year time frame, and invest — not so much in a specific product or technology, but we're looking more at how we can nurture the core competencies from which we get our economic growth. We're trying to identify what we have to be really good at, and we're making sure some of our R&D dollars go there."

What Techware engineers are working on now can still be traced directly back to the original products, and simply represent improvements on them. "There've been massive software updates, of course, and changes in the hardware, but fundamentally the concept is the same," says McMahon. "New generation platforms, faster computing, adding connectivity to other computers, graphics, packaging, ways to provide good service in a place like Korea — we've developed all kinds of methodologies. But it all types back to what we started with."

Today, Techware's business is evenly split between capital equipment manufacturers and end-users of control solutions who are also large research, development and production facilities, such as Xerox, IBM, Motorola and Intel. Interaction with these end-users, says McMahon, fosters Techware's understanding of what kinds of new technologies will be required in the immediate and long-term future. In other words, the strategy for now is to let the market drive the technology.

One feature that McMahon thinks makes Techware unique in its market is its "generic" approach to equipment control. Techware's equipment features modules that allow customers to choose the components that best suit their needs. "When integrated into an entire product line or facility, Techware's generic controllers provide a common user interface and rapid development environment, thus greatly reducing development and support costs," says McMahon.

"Operating in both research and production capacities, Techware controllers' flexibility and ease of use result in increasing up-time, more reliable processing and standardization of operations."

Some of Techware's "new generation" products include the Techware II Process Equipment Controller, the TC-111 Tool Controller, the Cluster Module Controller and the CONTROL-Vision Integrated Control Software.

Setbacks and Mistakes

Around 1985, Techware designed a second product — ABC, the automatic battery cycler — that had very little to do with the original control system, and it was ultimately discontinued when the company outgrew the potential size of that market. This was virtually the only time the company directed its efforts off course, and fortunately there were no negative repercussions, because they stopped making the product as soon as it became apparent that the market was too small.

Diversions from the main product line can sap energy and resources.

"The product was designed for cycling rechargeable batteries, the idea being that by controllably discharging and recharging them you could characterize the battery and find out how well they lasted under different discharge and recharge conditions," says McMahon. "This instrument was used by a variety of research people who were looking at new battery technologies."

It wasn't completely disconnected from control systems technology. The two products shared some core software, and both involved real-time, multitasking upgrading systems, some measurements and some control. "We got into it because of a company here, a spinoff of UBC physics, that set up to commercialize a certain type of rechargeable battery," says McMahon. "Since we knew those people from our grad student days, we were able to say, 'You have this need, we have this technology, maybe we can build it into something that'll be useful to you if you'll buy 10 of them,' and eventually they did. It never really went anywhere from there."

The recession of the late 1980s forced McMahon to lay off half his staff, and he concedes that if Techware had had more staff with significant management experience, such drastic measures might have been avoidable. He maintains, however, that cutbacks had to come in some form because of how closely tied Techware is to the capital equipment market place, which was badly hit by the recession.

"The number of employees grew gradually," McMahon says. "At one time, before the recession, we were at 35 people. We had serious layoffs during the recession, and by 1990–91 we were down to 14 people.

"Why? In retrospect, we probably could have found ways to manage ourselves through the recession if we had more experience. But we didn't. On the other hand, we're delivering capital equipment into the capital equipment market place, which is then used to produce commodities. When they start losing 10 percent of sales to a recession, the first thing they'll stop buying is capital equipment. We're a third party supplier to capital equipment, so they look at the numbers and decide not to contract out as much, to do more in-house production instead. The contractors get laid off first — that's us. Economic trends can be pretty hard on us. That means the downs are more down, the ups are more up. There's a bit of instability."

Risks and Choices

The scariest close call that McMahon can remember happened right in the middle of the recession, when the Bank of Montreal threatened to foreclose on their $400,000 debt.

"We were completely unprepared for them to call the loan like that," says McMahon. "All along, the reason we were on the financial path we were on was that we knew we had made a major sale to a large manufacturing company. The deal was just about to close, we'd been putting a lot of work and a lot of investment into it. We knew it was just weeks away." Just before the final due date for the loan repayment, the sale came through. "And we got lots of money from it, and paid off the loan, and within a few months had quite a bit of money to deposit in the bank," says McMahon.

By then, the staff had dropped to 14 people from 35. "We had definitely point by point been cutting back," says McMahon. If the time it took to make the sale had stretched on much longer, he says, it really would have been up to the bank. "Even before the sale came through there was a point where they were really monitoring us, watching us manage what resources we did have, seeing the layoff notices being issued before and after they called the loan."

In retrospect, McMahon says, Techware had probably been expanding too rapidly just before the recession. He attributes many of the money problems they experienced during that phase to the move they made in 1989 from a crowded office in downtown Vancouver — where 25 people had been working in a 1,200-square-foot office space — to their current, roomy location in Richmond. "We were getting nice sales, and we moved into this building. Then we got into debt with the Bank of Montreal to pay for some renovations, and we probably just weren't quite conservative enough.

An understanding of the economic environment is essential.

"Everything was going fine, but when the bottom dropped out of the market, we couldn't get out of the debt we were in fast enough, for legitimate reasons."

It's a problem Techware will face again in another year or two, says McMahon, and there are still no easy answers.

"One of the problems with moving when you're an expanding company is you're totally reliant on a projection of how big you're going to be," he says. "You need to pay for something that's too big when you move in because you're hoping to grow — for instance, we had been growing at a rate of 100 percent per year. It's hardly sensible to move into a place that's only big enough for the next year, but how many years ahead do you plan for?

"We're faced with that question now. This place is big enough, but looking at the growth we're doing now, we're going to have to move again. We don't want to get in over our heads. I directly correlate that move and doing all the tenant improvements to getting into debt with the bank and eventually being called on it."

The Outlook for the Future

McMahon's long-term vision for Techware is to make it "the number one supplier of control systems to the top 10 semi-conductor companies in the world — IBM, Motorola, Intel, Samsung, et cetera. We want to be their main source of controls in their equipment." He admits that goal is ambitious since those companies still favour producing that kind of equipment in-house. "It's a long-term guiding light more than anything else," he says. "We've only got 1 percent of the market so far."

Techware has traditionally done most of its business in North America, which still makes up about 60 percent of its market. Europe is good for another 30 percent, and Asia for the remaining 10 percent. "The recession seems to be particularly bad in Europe, but Korea is really taking off," McMahon says. In North America, the United States is where Techware does most of its business. "Canada doesn't have much thin film industry," says McMahon. "There's Northern Telecom and a few smaller companies, but Canada has never amounted to more than a few percent of our market."

If its market starts to get more responsive, Techware will be in a position to increase its revenue in the not-so-distant future. "Right now the market we're in is much bigger than we are, and there is a trend towards a desire to have third party controls," says McMahon. "People are looking at us who weren't before. So as long as those sorts of trends keep going, we'd like to follow that. We're in a really good position and it makes sense to capitalize on that trend."

On the other hand, he says, they're also interested in keeping tabs on what's happening in the broader market, if for no other reason than to know what might spring up out of nowhere as an alternate solution from some other industry. "Something could just happen to fit our industry, could come in with a lot of force and take over," says McMahon. "In that case, we want to have a broader perspective. Whether or not we actually sell in the broader based market is a good question."

For the future, the driving technological force will continue to be the semi-conductor industry. Because of this, ongoing commitment to research and development is crucial to the evolution of Techware's product line. Techware will continue to focus on providing advanced control features in order to meet customers' evolving needs and technological demands.

And customer feedback will continue to be important to product development initiatives. "Additional hardware components and software features are continually being developed to provide customers with new process control technologies that will improve the quality of processing for a variety of thin film applications while adapting to evolving equipment standards," says McMahon.

The Competition

McMahon says Techware's product was always unique, and in some ways still is. "We essentially invented the concept of a third-party control system," he says. In-house control groups that build controllers for various product lines are the main competition. "But we're really the first to come along and say, 'Let's do this for a broad category of equipment types and essentially be an off-the-shelf supplier of controls,' just as they buy pumps and power supplies off the shelves. We're still the only people doing it as a stand-alone business."

"The market is worth something in the hundreds of millions. We're about a $4 million company," he says. "But we see a lot of growth potential, because the industry is starting to wake up to the idea that maybe each capital equipment manufacturer should not be building stuff they can buy off the shelf. Capital equipment manufacturers are starting to look at things like what their core competencies are, and they're things like processes, like the vacuum equipment. Is it software? No, not really; but they have a 20-person software team. Why? It's an empire. It has grown within them, they thought they needed it, and they're starting to look at whether there's another way they can buy it without reinventing it."

The trend in the market place is opening up to third party suppliers, according to McMahon, and that puts his company in a very promising position.

Management Strategy

McMahon, who dresses casually for work and likes to foster creativity among his staff, is no big fan of the business plan. "Ultimately in a rapid-pace technology business, it seems that by the time the ink is dry you're already realizing how out of date it is," he says. "So we're just trying to focus on things like vision, mission statement, and core competencies." His strategy is to deal with today's problems and make sure the company's investments are pushing it in the direction that the long-term vision says it should go.

> Some firms seem capable of operating in a looser management mode.

McMahon founded the company when he had no prior management experience himself, and successfully grew his business to where it is now with a partner who also had no experience — and that explains why he doesn't think it would be exceptionally useful to hire anyone specifically for their managerial aptitude. Not only that, but he's pretty sure that sort of person wouldn't really fit in anyway.

"We've made quite a bit of use of consultants, especially in the training side, to help inject some management savvy and academic management ideas into our heads, but we haven't had any great success in bringing in 'professional' managers to the company," he says. "It's a cultural difference, perhaps. We're still sort of free-wheeling. Informality is an important value of ours, along with dedication and innovation.

"We're ultimately pretty hardworking people — which isn't to say professional managers aren't, but there's a kind of a 'roll up the sleeves and just get it done' attitude here. Those of us who started in the 'doing' end of things have been educated to lift our heads up, look around, see the future and decide how to modify the company to get greater success, without necessarily bringing someone in to tell us how to do it full-time."

Something McMahon says he thinks about often is the model of technology transfer out of universities and how to make that happen.

> International alliances have become a normal way of doing business in knowledge-based industries.

Techware is a member of the University of Michigan's Semi-Conductor Research Corporation, a consortium of semi-conductor industries that looks at how technology transfer can best be achieved. Techware became the SRC's first Canadian member in 1990. "It funds all this research at universities that so often seems to die on the table," says McMahon. "The kind of stuff that's close enough to commercial that it's no longer thesis material, but it's also nowhere near ready to be actually purchased and imported by a company. I see a lot of opportunity for grad students to essentially follow the model I did, which is to say, 'Look, nobody else is going to market this. I know this inside and out, because I created it as my thesis, and what a wonderful entrepreneurial opportunity.'"

All it would take, according to McMahon, is a little support from both universities and the government, who should see this as a way to create small business, to further entrepreneurial high-tech, and to foster that environment.

"I would hope that university students in that situation wouldn't just say, 'Look, the only model of a career I have is to go work for IBM or one of the big giants.' I hope they realize that there is opportunity to commercialize the work they've done at universities, in many cases.

"It's risky, but it can be a lot of fun, too."

7

Instantel Inc.: Getting a Second Wind

Instantel is a small company that upset major US firms by applying leading-edge instrumentation technology to their traditional market. However, having captured more than 40 percent of the world market, rapidly it found itself at the top of the S curve with that product line. It has introduced two new product lines that it expects will generate growth rates of 50 percent annually for the next five years.

Instantel, a small, privately owned company, was launched in 1982 as a collaboration between a pair of engineers, one from the high-tech industry and the other with a mining background. The company, which is located in an industrial park in Kanata, Ontario, and now employs 35 people, started off manufacturing seismographic instruments to monitor blasting levels at construction sites and quarries.

Today Instantel is the only Canadian company that manufactures instruments for monitoring blasting operations in mining, quarrying and construction industries. Its chief competitor is an American company, GeoSonics, of Warrendale, Pennsylvania. Recently, Instantel acquired two other types of completely unrelated technologies from local companies. Now the company is divided into three sections according to product type and targeted market: instrumentation, tracking devices and detection systems.

Instrumentation products — such as portable, low-power recording devices — are designed to measure noise levels of activities such as construction site blasting by monitoring ground vibration and air pressure. Such noise, when its intensity is not maintained at certain levels through regulation or monitoring, can cause damage to windows, foundations, sewers, power lines, and pipeways. Instantel's seismographic software is copyrighted, but the company buys its hardware elsewhere.

Tracking devices rely on radio frequency technology and are used by hospitals and similar chronic care facilities to monitor patients, allowing them more freedom of movement: instead of being physically restrained or sedated, patients who require monitoring wear tracking devices as wrist watches, which alert staff if the patient leaves the facility.

Detection systems also work on radio frequency technology and are designed to monitor property, such as car lots or industrial sites, by detecting the presence of intruders. These devices are still in the development stages.

Brian Martin, the current president, says he expects the young company to grow at a rate of 50 percent per year for the next five years — a rate he admits is somewhat risky. For him, it's also exciting: he has an extensive background in high-tech, but has never had complete control over a company, or managed a start-up business. He started out with a degree in electrical engineering from the University of Waterloo in 1975, and from there went to work at Computing Devices in Ottawa as an engineer. After three years, he moved on to a position as a design engineer at Bell Northern Research. His next move was to Hyperion, a computer firm, where he was director of engineering; and most recently, he was director of product management at Mitel for two years.

"This is my first mission in running the whole shooting works," he says, looking confident. "It's my first kick at the can."

Instantel was launched after one of its founders, Denzil Doyle, heard a mining engineer complaining about the technology that monitored blasting at construction sites and other similar projects. For instance, if a dispute erupted over blast levels at such a site, the whole project would sometimes end up shelved for as long as a month following the initial explosion while lab results were analyzed. At the very least, with no way to monitor blasting levels, construction companies would continue to operate while results were analyzed, never knowing if they were contravening blasting regulations. These delays almost always resulted in increased costs — rental equipment could sit unused, climate change would make the job more difficult, and employees would have to be paid longer, for example. There was clearly a need for better, quicker technology to measure this kind of noise. It seemed obvious that any company that could design such a product would almost certainly find itself becoming indispensable to industries involved in blasting activity.

Brian Martin won't reveal sales figures because of his company's intense rivalry with its competitors south of the border. "Anything you can find out about the other guy that they don't know about you is an advantage," he says. "Let's just say this: we've been profitable for the last seven years, and we expect to grow by 40 to 50 percent per year over the next five."

Instantel sustained a few years of losses while it was still learning to fly. The two founders, says Martin, had other business interests to pursue, and never intended to manage the company themselves to begin with. So almost as soon as they started, they brought in a management team that would run the company, never really playing an operational role in it themselves. Martin was imported in 1986 because the original management team wasn't working out — there were problems with direction, and the company "wasn't sustaining itself," says Martin. "There were a number of mistakes made, including some technological problems."

> Lack of direction can lead to dissipation of energy and resources.

For one thing, the company was making three different models to monitor blasting — and each featured a completely different microprocessor, hardware design, and package. "There was a very strong negative impact on costs, in research and development, and on manufacturing especially," says Martin. "We were paying for everything three times over. There were also reliability problems."

For instance, it was someone's idea that the product should fit under an airplane seat. Martin says that's an example of an idea that needed someone else to step back, take a good long look at it, and examine its flaws. "The idea probably came from laptop computers, which were becoming more and more popular," says Martin. "We ended up with this thing that was incredibly costly, incredibly heavy, and basically just didn't work that well. It compromised the entire packaging design to the point that the thing weighed very close to 40 pounds."

The other models, says Martin, included one "non-starter" not worth describing, and a second product with packaging and reliability problems that wasn't rugged enough to function in mining environments. Essentially, says Martin, the problem was that "nobody took a step back and said, 'Wait a minute. What the hell are we doing here?'"

In spite of all this, because of the concept on which the company was founded, sales didn't suffer too much, says Martin. "Up to this time, seismographs had recorded the data on tape, and that went to a lab for analysis. It was an inconvenient, expensive process."

When Martin joined Instantel during 1986 and started turning things around, one of the first changes he made was with the technology: he decided to make the three existing models more similar. Now, production is streamlined so that the devices are virtually the same in design and manufacture. "For the first 90 percent of the way, they're the same," says Martin. "Then in the last 10 percent you can decide which model you want it to be." The result is greater flexibility and huge savings in development and manufacturing costs.

Instantel also learned to refine its technology by shaping it to suit user requirements. This is a strategy upon which the company has relied successfully, and continues to use. In the beginning, customers defined their monitoring equipment requirements as: quicker, on-site analysis; rugged equipment that could operate in harsh environments; and simple, user-friendly operations that would be easily understood by people with little education and no computer training. Instantel responded with a 20-kg machine that featured microprocessing and printing technology, a plain alphabetic keyboard, and intuitive operations that provided users with an instant, printed hard copy of the results. Essentially, all innovations in seismographic technology have been improvements to existing technology. All Instantel has to do is develop the software.

History and Milestones

Although Martin declines to discuss specific figures for sales and revenues, he describes his company's 11-year growth curve like this: "There was growth in the beginning; it levelled off for two or three years; then it dropped slightly, because we had a big contract one year that we knew wouldn't happen in the next year. This year there's been a 50 percent increase in growth over last year, and that growth rate should continue for the next few years."

The company ran its affairs smoothly in the early years, says Martin, until "the fellow who was running the company at the time decided that a really neat idea would be to get into the hairdressing business — where you take pictures of people and use a computer to superimpose hairdos on them prior to the haircut," says Martin. "That's the direction he wanted to take the company. But you've got to have a mindset that says, 'Look: This is what the company does, this is what we're working on.' That doesn't mean you shouldn't look at other ideas, but this one was so 180-degrees different that it could have put the company in tremendous stress." That was when Martin was brought in.

Martin made an entrance by initiating some personnel changes. "I brought in a very key engineering manager, and a finance guy," he says. Of the original twelve, there are only four in the company now. "We redeveloped the three models of the seismograph with one common hardware design and one package, and the differentiation is only in software features or minor packaging things." Commonality of components is a more cost-efficient way to do business.

> Working out a product migration strategy is key.

Forecasting, says Martin, is a very difficult thing to do in terms of sales — and even more difficult with three different models. "The redevelopment helped us to forecast as best we could, but still allowed us to make last-minute changes. In other words, we weren't stuck with Model A when we had an order for Model B. All three sold to the same markets. We also dropped the costs in half for manufacturing."

As soon as he introduced the new seismograph line, he raised the selling price. A clever money-making scheme? Not really, says Martin. "That may sound a little strange, but a number of things had to be done. First of all, you drop the cost. Then if the price stays the same, it would appear that you make more money — and in fact you do, on a per machine basis. But you've also got a lot of R&D to pay for as well. So we looked at the market, looked at where we were in the market, looked at the competition, then announced a new model and raised the price about 5 percent."

Martin says he wanted to clear out inventory of the old product as quickly as he could, and also wanted to get the new product on the market with all possible speed. "So we said, 'This'll be the price in three months from now, so get your orders in now at the old price of the old machine if you want the new machine,'" says Martin. "That got the machine out there quicker so we could get towards the next step." The new machine took Instantel engineers a year to design. It weighed about 15 kg, had fewer components, and was significantly more reliable.

In 1991, Instantel again introduced a completely new machine about half the size of its predecessor. "We exploited as much as we could on the first technology," says Martin, "but there were new components coming out that were much more efficient in terms of battery consumption. So we could get rid of one of the batteries, and that saved six pounds right there. And because the components were getting smaller — we started using surface mount components — we managed to reduce the size by half for all three models."

At the end of 1992, Instantel introduced Mini Mate, a seismographic device weighing less than 2 kg. It had fewer features than its predecessors, but sold for half the price. "It was an excellent deal for people who couldn't quite afford the $6,000 machine, but had to start monitoring," says Martin. "These things are like seatbelts. Nobody wants to buy them, but they have to."

By the time Instantel introduced its second machine in 1991, they had so infiltrated the United States market that each of their two main competitors were showing strong defensive reactions to their presence. Dallas Instruments, the company they were most afraid of, stopped manufacturing seismographs for all intents and purposes, letting that part of their business die off.

GeoSonics, a company about the same size of Instantel, from Warrendale, Pennsylvania, was so astonished at Instantel's speedy follow-up that it became convinced Instantel must be receiving unfair government aid — or subsidies, or both — and filed a complaint with the United States International Trade Commission, insisting that Instantel and its 26 employees were "hurting the entire industry south of the border." GeoSonics had been manufacturing a competing machine that had enjoyed a four-year lead on Instantel ("while we were making all those mistakes," says Martin). They were shocked that another company could launch two completely new machines in a three-year period. Their allegations prompted a further investigation into the matter by the US Commerce Department in March 1992. "They said we were being subsidized, otherwise we couldn't have beat them," says Martin. "We had the United States Commerce Department up here going through our books, and I had to go and testify in Washington."

> The US Department of Commerce is a ready ally of American firms.

The department applied a tariff beginning at the end of March 1992 on Instantel's seismograph imports that added up to about a dollar per unit — 0.02 percent. ("It's nothing, they don't even bother collecting it," Martin said at the time.) Instantel "won" the case, says Martin, proving that its product was designed and manufactured without government support, and was simply the result of talent, innovation and a clever marketing strategy. "To make a long story short," he says, "GeoSonics withdrew their petition. Behind the lines, what that said was that they had lost; they knew they had lost. They left the battlefield mortally wounded, you could say."

That positive outcome, and Instantel's subsequent ongoing success in the American market, convinced the company that the tight, lean company structure they had been working with really was the best way to go — the dispute proved that it was the fastest way to respond to rapid changes in a continually shifting market.

Shortly after Instantel's win, GeoSonics's main competition, Vibra-Tech — another vibration consulting company — decided they didn't want to make machines any more. "They wanted our machines," says Martin, "so now we private label our machines for them."

In other words, Instantel turned that competition into a customer. "They happen to be bitter enemies with the people who launched the countervail," adds Martin. "So we were selling arms to the other side, to speak."

Around 1988 Martin and his staff started looking at other ways to build the company. They had quickly established themselves as the industry leaders in the seismograph market, but although it was a profitable market, it wasn't growing. So they did something that took the US market entirely by surprise — they signed up Atlas, a "powder" company that made explosives, as a national distributor. "They were selling to the exact same people we wanted to get to," says Martin. "So we made a deal with them to distribute our seismograph. This really upset our competition. They were saying things like, we couldn't do that because it was a conflict of interest."

As the market possibilities saturate, a strategy to move to a new S curve is required.

Even this deal could only take the company so much further. "There's always a certain portion of the market you're not going to get," explains Martin. "It's easy to get the first 10 percent of the market, reasonably easy to get the next 20, 30, 40 percent; but once you get beyond that, you hit this linear curve where you spend a tremendous amount of money just to gain a small further percentage of the market. We didn't want to do that." So Martin started looking for other opportunities.

He began by considering earthquake monitors — but the market was too crowded, and Martin couldn't see a way that would distinguish Instantel from established competitors. "It would have required a quantum leap in performance and/or price," he says. Next, he looked at monitoring machinery in industrial complexes. He found the same problems. Finally, he decided the best solution was to acquire a completely new technology.

"So three years ago we purchased two completely unrelated technologies," says Martin. "The first was a radio frequency identification technology from a local company, Total Alert. They were making miniaturized radio frequency transmitters that could be worn by people for monitoring in nursing homes and chronic care facilities." These allow people complete freedom within the facilities, so they don't have to be physically restrained or sedated, but if they try to leave the facilities, an authority will be alerted. "We acquired that technology for a few reasons," says Martin. "One was that the RF technology had a much broader base of applications than seismographs. It's more than just the health care market — it's article tracking, assets management, in-house arrest, electronic license plates, et cetera.

"We did a technical audit of their product over six months, then threw it out and redeveloped it entirely. They had numerous problems — cultural, technical, performance. They were trying to run before they walked. Basically, they ran out of money in the process. Customers were upset with the products because they never worked. There were lawsuits going on." Instantel began to improve the technology, working out the kinks and exploring the possibilities.

About six months later Instantel acquired a second technology that was in a much earlier stage. "The RF identification technology is basically a security product," says Martin. "This new one was also a security product, but instead of securing an asset or a person, it would secure property — for instance, the perimeter to an industrial site or a car lot. The person who had the idea didn't want to start his own company. So we bought the rights to the idea, the patents, and created the technology. It was risky, it was more fundamental research than we'd ever done before ... the whole thing could fail."

Martin says Instantel expects to have first commercial sales of that product shortly. "The market for it is tremendous. We knew that in advance."

Instantel introduced its tracking devices in 1992. He says they've been doing very well, but could still do a lot better. "There were some initial technological problems," he says. "Our main problem was distribution. We're slowly sorting out those problems."

Instantel's sales doubled in 1993. "We expect within two years to have our fair share of the market. We've concentrated mostly on North America until now, but there's significant interest from countries like Japan. We do want to go international." About 80 percent of Instantel's revenue is generated outside of Canada.

The most recently acquired technology — detection systems to monitor property — is currently being tested at Instantel's headquarters in Kanata. "We have three systems up and running right now, we're doing tests on them," Martin says. "It's an idea that is so simple that everybody can understand it when I describe it to them, but going from that simple idea to actually creating a product that works is a lot of effort, and that's where we've been putting the development dollars. We spend an average of 20 percent of our revenue on R&D."

> A mix of funding instruments nearly always comes into play in the start-up phase.

Martin says Instantel received little government support when it was first trying to get off the ground. "There was an equity infusion from the founders and shareholders," he says. "There was also money injected into the company on a debenture basis from the Ontario Centre for Resource and Mining Technology to develop technology for the mining community; that was paid off by 1988." The company also received funding from the Industrial Research Assistance Program.

Instantel divides its main customers into three distinct categories. One includes hospitals, nursing homes, and long-term care facilities, for the tracking devices; another includes industrial and commercial people with outdoor assets that need protection from vandalism, or people who need to protect the public from their potentially dangerous assets, for the detection systems; the third category consists of quarries, mines, underground mines, blasting and vibration consultants, government, municipalities, major construction and contracting firms for their seismographs.

The company has grown gradually but steadily, from 12 to 35 employees. Its products now sell in about 45 countries around the world. "We're pretty sure we make more of these [seismographic] machines than anybody else," Martin says. Their market share is tough to calculate, he says, because analysts rarely bother to examine it and figure out what it's worth — it's a small, specialized niche market. "But if you pressed me for a number, I'd have to say we have 40 to 50 percent of the market worldwide, and in Canada certainly close to 98 percent. The overall market [for seismographs] is worth about $5 million. It's not growing, though, and that's why we got into the other businesses." They have recently been involved in such high-profile contracts as the expansion of the Hong Kong airport, Boston Harbour dredging and Korean transportation tunnels.

Instantel's staff is divided evenly among manufacturing, R&D, sales and marketing. The company still does all of its marketing, manufacturing, research and development at its single location in Kanata, and its response time is therefore quite rapid. For instance, a few years ago the company decided to incorporate a competitor's feature of seismic analysis into its own product, and was able to do so in only three weeks, whereas in the reverse situation, the competitor would have lagged sadly behind, as its manufacturing — and some of its design — facilities were contracted out to other firms.

Risks and Choices

As a relatively young company just beginning to take off, Instantel has few setbacks to report and no major mistakes to regret. If there are large errors to be made, they will probably happen in the next few years. The closest brush the company has had with a setback was in the early part of its history, when it manufactured the three completely different seismographs. Had someone realized the inefficiency of this system earlier, it's possible that greater growth and profits might have come sooner. Still, Instantel was lucky enough to have a market so hungry for its products that sales were still strong, offsetting some of the inordinately large R&D and manufacturing costs.

In its first few years, Instantel's biggest concern was that one of their big American competitors would "cotton on" to what they were doing, says Martin. "We figured, if they decide to do it themselves, we're dead in the water. They're bigger than us, they have the market share, they have the penetration. They had everything."

They had everything but Instantel's secret, and the fact that they didn't catch on until the last minute gave Instantel time to gain the foothold it needed to be successful. "However," says Martin, "the foothold was very tenuous because of all these product problems we were having."

> Established companies are often blind-sided by technological up-starts.

In the past, says Martin, "the major risk was being up against the major players and never knowing when or whether they'd get smart and kick you out of the market if they had the chance.

"In the tracking devices, the major risk is making sure you've got the marketing and distribution in place, and sometimes that's beyond your control. In a start-up company, money is risky. You need good cash flow. If you don't have cash, you don't have the necessary freedom to make decisions."

"There's a strength and a weakness here in that growth is not coming from a single division."

The Outlook for the Future

Instantel's future holds probable strategic alliances with firms that have similar customers and markets. For instance, the company is involved in negotiations with other geotechnical companies that it would like to use as distributors. "Any strategic alliances we form will be largely in the marketing and distribution areas," says Martin. "We're very good at everything else. We're in discussions right now with two, getting them to distribute our product."

> In technology companies, those who listen to the client base clearly benefit.

Instantel is working on an upgrading scheme that would allow customers to install new features in their existing machines for a reasonable cost. The idea is to foster long-term relationships with loyal customers, allowing for continued feedback and revenues. The whole plan will require ongoing development of topnotch, innovative products to satisfy changing market demand. The company will try to keep its innovations market-driven, listening to what customers have to say about existing products and how they could be improved. In the past, for instance, customer suggestions have led to longer battery life, waterproofing, smaller, lighter equipment and easier installation of printer paper. This add-on philosophy is likely to continue as long as the memory functions of the computer hardware are not dramatically improved, in which case the whole product would have to be changed.

Martin expects spending on R&D to drop as a percentage of revenues, although the absolute dollar value will remain the same. "We'll spend more on marketing and distribution. That's the trade-off we'll have to make."

On the subject of going public, Martin seems indifferent. "The market looks good, but we just don't need to right now," he says. "There's been almost a herd mentality in the last year or so about going public. You're only going to get one kick at the can in the public market, and we want it to be the right time."

The Competition

When Instantel first broke into the instrumentation devices market, they knew they would be competing in a field full of experienced players who already had reputation, resources and size on their side. The only way they could hope to survive was to offer a product that would either be a radical improvement over existing technology, or be much cheaper. Instantel's product was both: it was much faster to use, and considerably cheaper in long-term use. Instantel's price per unit seems high compared to its competitors' fees — $6,000, compared to a range of $500 to $1,000 — until you consider how often a company would have to pay that $1,000.

In that sense, Instantel's product was revolutionary in its market. The company also had a considerable lead time; competitors became alarmed about Instantel's new technology far too late to do anything about it. Competitors tried, for a while, to do some damage repair by coming up with new twists on their own existing technology, but failed to produce any innovative application that could challenge either the convenience or long-term cost efficiency of Instantel's product.

A leading-edge product can outdistance the competition.

Management Strategy

Martin says Instantel's formula for success is built around "a healthy dose of common sense in what you're doing.

"Be prepared," he says, "to take a step back and ask yourself: Does this make sense? It's easy to get into a rut if you don't ask yourself *why* you're doing something. I'm constantly challenging people here: Why are you doing something this way? The right answer is not 'Because that's how we've always done it.' The answer has to involve some common sense."

Feedback from customers is also crucial, and has played a key role in shaping Instantel's products. Also, says Martin, "Don't just assume you've got a better mousetrap — don't ignore the market. Know in advance who's going to buy your product. Make sure you've got the distribution and the marketing. And you've got to be able to justify spending — you can't spend 100 dollars on R&D and 10 cents on marketing. It isn't going to work."

Diffusion of a product in the market usually takes much longer than expected.

Entering a market, he cautions, also takes much longer than most people think. "Don't underestimate the length of time it takes to set up distribution," says Martin. "Don't underestimate the value of word of mouth. It doesn't matter how good your product is — there's still a finite amount of time that will have to go by before people use your product, say, 'Hey, this is really good,' and start telling other people about it. There's not much you can do to shorten that time."

Next to common sense, Martin emphasizes the value of word of mouth in sales. "We've got a deal with a major American company about national distribution, and that's because they went out in the market place and asked customers what systems they used. Our name came up.

"In another instance, a customer called us and said. 'I want info on your product.' We asked how they found out about us. They said, 'It was very easy: We've got a list of all the people that make this product. We called them all. They all said their product is the best. We said, "Well, if your product is the best, whose is the second best?" They all said Instantel.'"

Now that Martin has start-up down to a science, he will have to devise ways to manage the rapid growth he's predicting for Instantel in the next few years. "We've grown 50 percent this year from last year. We haven't changed much," he says. "But we're not going to be able to get away with that in the future. We've put a lot of infrastructures in place. We're a very curious company in that we're running three businesses. We have a common R&D group, we have a common manufacturing group, we have a common admin. group. But we have three business divisions operated by three people. Clearly there will be resource conflicts. They all want R&D, they all want manufacturing. There will be a fairly healthy contention for those resources."

He plans to play things partly by ear, fielding problems as they come up. "We know it's coming. I don't know how we're going to manage the growth," he admits. "Our five-year plan, put together recently, shows significant growth, averaging 40 to 50 percent over the next five years or so. That is approaching the dangerous level, in the sense that the left hand may not know what the right hand is doing. We have been very careful not to put a lot of bureaucracy in the company that might have to be undone later. In that kind of growth you could quickly start losing track of things."

He will try to strike a delicate balance between the bureaucratic infrastructures that will be needed on an organizational level, and the creative, less professional entrepreneurial approach that has driven the company so far. Although the company is certain to require experienced individuals as senior managers in the near future, Martin is hesitant to go on any hiring binges just now, in the hopes of avoiding a multi-layered bureaucracy later.

He's also cautious about revealing any further secrets lest his competitors catch on to his strategies. "We're putting money on the bottom line now," he says. "Prior to the last seven years there were losses, but that's normal for a start-up company. When I had to go to Washington [in the GeoSonics case] I was forced to sit there and explain why my business was so successful, essentially giving the competition a lesson in how to run a business. Assessing a competitor's strength is an extremely valuable tool, especially if you know more about them than they know about you.

"So now," he says, looking smug, "we've adopted the philosophy that you walk softly and carry a big stick. It's not an original tactic, but it might just work."

8

Creo Products Inc.: A Seasoned Technology Manager Takes On a Mid-Life Challenge

Creo is the story of an experienced engineer with an MBA who has a vision of how to manage creative people in a leading-edge technological area.

When Creo Products Inc. was first founded in 1983, its dozen or so employees worked out of a small, rented office in Discovery Park, a research park in Burnaby, BC. A decade later, with 120 employees and more than $10 million in sales, Creo owns the entire building and now leases parts of it out to small start-up companies like the one it used to be.

Founded by Ken Spencer and his partner, Dan Gelbart, Creo — whose name is Latin for "I create" and Spanish for "I believe" — designs, develops, manufactures and markets high-precision electro-optical products based on patented technology. The company's two main product areas are optical data storage and photoplotting. Its two general market products are an optical tape recorder for large volume storage, and a precision photoplotter for printed circuit boards.

In the first few years of its operation, Creo focused on researching and developing high-precision electro-optical products, and supported itself through custom design, becoming involved in the development of an optical inspection machine for a California-based client. As the only contender in a market niche that it created itself, Creo is now poised at the beginning of a fast track to major success.

Spencer, still president, is an electrical engineer with an MBA and a PhD in electrical engineering from the University of British Columbia who has made entrepreneurship his "life's work." His first job was with MacDonald Dettwiler and Associates (MDA) in Richmond, BC in 1971. He joined the company when it could still count its staff on two hands, and left after nine years during which time the company grew to almost 300. He spent the last three years as vice-president and general manager.

He left the company in 1980 to sail around the world with his family. "I had my mid-life crisis on the Caribbean instead of in the rain," he says. In 1982 he came back. What he really wanted to do was start his own company, but there were two strikes against him: there was a recession in full swing, and he lacked sufficient funds. "It wasn't a good time to start," he says. "So I went to work for Glenayre Electronics."

He worked at Glenayre as general manager for about a year and a half, raising the necessary capital to start Creo. Looking around for a partner, he chose Dan Gelbart, an electrical engineer who had also worked at MDA as principal engineer. Gelbart had emigrated to Canada from Israel in 1973, and had expertise in precision mechanics and optics. Eager for an opportunity to have more control over what products he was developing, Gelbart took Spencer up on his offer. "He was the techie and I was the business type," says Spencer.

Creo became an example of a company led by a businessman on a mission — first to find a market, and then to develop a unique product that would become a runaway success.

Spencer says he was motivated to start his own company for several reasons. "If you're smart and successful, you want to share in the rewards," he says. "I had a lot of ideas about how to manage a company, some of them quite different than traditional management, and I wanted to try them out, but working at other companies, I never could."

Now he is putting those ideas to the test at Creo, and they seem to be passing with flying colours. "Some of them are ideas I had 10, 15, maybe 20 years ago that are sort of catching on now — the new term is 'empowering people,'" says Spencer. "Everybody at Creo can make decisions. We don't have a lot of managers making decisions about what affects other people."

> Many technology firms prefer a "flat" management organization to encourage creativity.

Creo's key products — optical tape recorders — are designed to hold a terabyte, or one trillion bytes, of information. The technology uses 32 lasers to fire data onto one side of the plastic optical tape (which is coated on one side with a metallic substance) at extremely high density. The information is stored in the dimples of the metal. That's how a single reel of 35-mm wide optical tape holds a terabyte — the equivalent of a billion sheets of typewritten paper, 1,600 compact discs or 5,000 conventional computer tapes. The potential customer for this product, says Spencer, is anyone who needs to store vast amounts of information; possible applications include seismic data logging, satellite image-processing, medical x-ray imaging and document archiving. The average 500-bed hospital could store a year's worth of data on just one of Creo's reels, making storage not only space efficient, but economical in the long run.

The Canada Centre for Remote Sensing was one of Creo's first major customers. A division of the federal Department of National Resources, the centre records and stores image data beamed from earth observation satellites or gathered by aircraft. The volume of data to be stored is staggering, and until the arrival of Creo, the centre was using much more expensive and bulky magnetic tape for that purpose.

> Strategic alliances are becoming an increasingly valuable instrument for the development of a firm.

Over the years, Creo has managed to form several strategic alliances to considerable advantage. The optical tape used by Creo for its patented recorder, for instance, was developed by ICI Imagedata in the UK. At the time of the agreement, ICI stood to make no money on the deal, but went ahead with it anyway because it predicted great potential. Now it looks like ICI stands to profit from Creo's endeavours.

Then, thanks to a deal made with American high-tech giant Honeywell, Creo has also been able to sell its recorders to United States defence and intelligence markets that otherwise probably wouldn't look twice at a Canadian company of Creo's comparably small size. The recorders are sold through Honeywell's test instruments division in Denver, Colorado.

In 1991, Creo began to branch out in a new direction when it won a major contract with a billion-dollar Japanese company called Dainippon Screen Manufacturing Co. Ltd. to build engines for its photoplotters. Photoplotters in turn create printing plates from pieces of film for the graphic arts and semi-conductor industries. The main customers are printing plants; each plotter sells for about $300,000. "It's like laser printers," explains Spencer. "One company makes most of the engines. In plotters, there's a drum, 32 lasers, a carriage, and so on."

History and Milestones

Creo supported itself by doing research and development contract work for the first few years. "When we started the company we didn't have any money," says Spencer, "so I just started doing contract R&D for a company in California, developing machines for them, and that funded the company."

He was getting research contracts to finance Creo's ongoing research on its own first product; but in the beginning, he admits, he wasn't sure exactly what that product would be. The one thing he knew for sure was what he *didn't* want to do: remain an R&D company. But doing preliminary research and development to start bought them some time to look around and make a few decisions. Within about three months, they had settled on a product: the optical tape recorder. A large part of the reason for this decision was Gelbart's background — most of his expertise as an engineer was in precision mechanics and optics, a highly specialized field in which he is considered to be exceptionally talented.

> Contract R&D, like consulting, is one of the entry points into the high-tech world.

Spencer says they had "no clear idea" about their potential market when they first made this decision. But before they made any solid commitment to the product, Spencer devoted a considerable amount of time to doing his homework. "I spent a lot of time talking to a lot of people, going to conferences, trying to understand the market," he says. "And then it took us so long to develop the product that the market place changed." But Spencer was still tracking it. "We knew it was changing," he says. "We saw the change." Fortunately for Creo, it was a shift that proved beneficial to them. The market was opening up; people were storing larger and larger amounts of data on magnetic tape. Ultimately, the creation of the optical storage tape was no stab in the dark. "We knew we could sell it," says Spencer.

But the development of the optical storage tape was hampered by numerous setbacks, and in the end it took Creo a considerably longer time to begin production and marketing than what Spencer had originally predicted. "It was a very difficult project," he says. "From the very beginning to the shipping stage, it took five or six years." Creo marketed the product aggressively through every possible route, says Spencer. "We made phone calls, went to trade shows," he says. "We do all of that — we go to the shows, we go to conferences, we have a direct sales force, distributors in Europe, and sales reps."

In 1986, after about three years of research and development, Creo's first big contract came from the federal government, which hired the company to complete the development of its optical tape recorder for use by the Canada Centre for Remote Sensing. The contract, which initially allowed for the delivery of four recorders, was eventually increased to an order for seven — two went to the Department of National Defence — for a total contract value of $3 million. "That contract funded our ongoing research and development," says Spencer. (The first optical tape recorder was finally delivered in May 1990.)

Meanwhile, in March 1988, Creo received a $325,000 grant from the Industrial Research Assistance Program to purchase computer-aided design equipment for producing and testing its optical tape recorders; later, more funding came in the form of $2 million from the Western Economic Diversification Fund and another $1.5 million from the venture capital arm of the Federal Business Development Bank, which still owns 12 percent of the privately owned company's shares.

> Tapping into all possible sources of government money is essential for survival.

Creo finally started supporting itself on its own product revenues around 1989, says Spencer. "We bootstrapped the company up until about then." At that time, when Creo decided to begin producing and marketing its optical storage product, it was able to do so thanks to funding from several sources: Creo had made some money of its own through custom design; it had received a total of $216,000 in grants from the BC Science Council and a $500,000 grant from the National Research Council; and it had been awarded the $3 million contract to supply the first of four optical tape recorders.

Then, in 1991, Creo won the contract with Dainippon to manufacture a plotter engine. "We took some of the technology we had developed, and some of our knowledge from other companies, and developed quite a different device: a photoplotter — a machine that plots large pieces of film very accurately," says Spencer. Dainippon paid $8 million for the research and first year's production, and Creo's executives anticipate potential for a much higher figure eventually as the Kyoto-based company continues to buy more of Creo's engines. That market could be worth between $4 and $8 million a year.

By the end of 1992, both the European Space Agency and the Australian centre for remote sensing had been signed as customers for the optical tape recorder. The BC Science Council named Creo winner of its 1992 gold medal for industrial innovation. And in the last of six different expansions, Creo took over the Discovery Park building in which it had first begun as a tenant 10 years earlier. The park, established by the provincial government in 1980, is designed as an incubator to support local high-tech companies that are strong on ideas and energy, but short on capital.

As of late 1992, Creo was manufacturing two recorders every month, each for about $250,000, and producing three different types of laser plotter engines for Dainippon. Their ultimate goal is to manufacture about 200 recorders per year.

The overall market for Creo's products, says Spencer, is worldwide: the chief customers for remote sensing storage are governments, and other customers include system integrators, mostly in the United States and Europe. "Zero percent of the market is in Canada, except for the Canadian government," says Spencer. "The market is growing, but we're struggling with that."

> The challenge for most technology firms is to get potential clients to move to a new S curve.

The problem will be convincing people to switch to optical storage. Because it's tough to forecast how quickly people will adapt to the newer, more costly technology, it's difficult to get much of a handle on the potential size of the market. "We're trying to replace microfilm, and there are some good arguments for switching to computer storage. But it's really hard to predict how quickly people are going to switch, and it's different in different industries," says Spencer. "The insurance industry is one we're studying right now. There's also banking, and so on, and they're not even sure themselves."

Also, Creo doesn't have a stranglehold on the technology; potential customers still have some other fairly attractive options to choose from. For this reason, Spencer says it's hard to predict what a reasonable market share objective will be for Creo. "We're not the only technology, so other things have to fall into place as well," Spencer says. "I know I should have all those answers. For some of our other products I do, but not for the optical tape recorder."

Currently, Creo does almost all of its manufacturing on site in Burnaby, and has a whole range of distribution channels, including direct selling and a sales force that sells mainly to system integrators.

And even though their main product has just recently begun to take off, Spencer is already looking ahead to new developments, including upgrades to the existing recorder. "We're starting to think about a new model now."

Setbacks and Mistakes

It's hard for Spencer to recall any near miss that caused the company much significant anguish. "I'm not one to panic," he says. As with any high-tech venture, there were hidden mines to be dodged everywhere and small close calls on a regular basis, but the closest thing to a disaster that he can remember involved a series of delays caused by suppliers that resulted in production of the optical storage tape being repeatedly suspended.

"We needed a chip, a special integrated circuit, and we went to a company that said they could do it in four months or something," says Spencer. After a year elapsed with no results, he finally cancelled that contract. "Then we went with another company who said they could do it in that same amount of time [four months] and it took them a year," he says. "So this thing took two and a half years to do, when we'd expected it to only take four to six months. That was a major setback.

Supplier linkages can be critical to success.

"So we had an optical tape recorder that could write data," says Spencer, "but we couldn't read it for about a year and a half while we were waiting for this technology. It had an impact on the time it took us to get to market."

Risks and Choices

Creo has an edge over competitors right now, says Spencer, and the big challenge for the future will be to keep it. "The challenge is, can we grow bigger and still keep attracting really good people? Can you grow bigger and maintain the responsiveness? We have to decide how to manage the growth in such a way that we don't lose what we have," he says.

Putting the firm on the stock market opens up a whole set of new concerns.

The two obvious ways to accomplish that goal are serious research and development that will allow Creo to keep a technological head start on competitors, and sufficient cash flow to support growth. For both of these reasons, among others, Spencer will find himself facing an important decision in the near future: To go public or not to go public.

In the past, he has been known to be fiercely opposed to going public. But Spencer appears to be softening his stance slightly; now, he says, it's not so much a question of being against the possibility so much as it's one of realizing what the risks are and weighing them against the potential gains. "I'm not opposed to it, but there are real risks, especially in the management of the company," he says. "Decisions tend to end up being short-term rather than long-term. People pay too much attention to quarterly earnings. Management attention turns from inside to outside. People start watching share prices, and not worrying enough about what's going on in the company and doing the right thing for the long term."

Still, he foresees a day, not too far off now, when going public will seem like a worthwhile trade-off and he will succumb to the pressure and the temptation.

"In the end, all of us who own shares want to get some money out of it," he says. "So we think we'll probably go public in the next couple of years. I'll have to eat my words."

The Outlook for the Future

Other plans for the future include strategic partnerships, a more broadly defined market, new products, international expansion, and a continuing high rate of research and development.

In partnerships, Creo will look for firms that would enable it to combine its technology with existing products. Spencer says he doesn't have anyone specific in mind as of yet.

Creo's short-term market includes the medical and satellite-imaging industries; in the long run, Creo hopes to target anyone concerned with document storage. Creo built some 30 optical tape recorders in 1991, and more than 60 in 1992 as the market developed. As noted earlier, however, it's tough to forecast when or how much that market will open up.

Photoplotters, on the other hand, are much more predictable. "We know a lot more about that market," says Spencer. "That's the graphic arts market, in which we have three different products — and some of them will be growing quite significantly over the next five years.

"We just announced last week a machine that we're working with that will be starting to sell next year. And that's a new market, in the sense that it writes a printing plate directly on a plotter. Right now, if you want to make a printing plate for an offset press, you make a piece of film and then you take the film and contact it on the printing plate," Spencer explains. "We, and a couple of other companies, are developing a plotter that will plot directly on the printing plate."

That's new, he says, so the market is starting at zero, but he expects it eventually to turn into about a $100 million market with some intense competition. "We're all just struggling to see who's going to get there first right now," says Spencer. "We expect to get about a 40 to 50 percent market share. We want it and we expect it."

For this reason, Spencer plans to install a sales force in Europe. "For the tape recorder, we're using reps," he says, "but for the new product we'll probably hire people."

The company currently spends about 20 percent of its revenues on research and development. That percentage has fluctuated over the years because the company is still relatively new. Still, says Spencer, "It's really high. In 1991 R&D was 50 percent of our revenues, but then we grew the sales. The next year it was $4 million out of $13 million. In the long run it's going to be about 20 percent. It's high." About 50 out of 125 staff are currently working in research and development.

> Initially R&D expenditures are very high to get the product "right." Over time these expenditures diminish relatively to marketing, quality assurance and other expenditures, but they still remain important.

The Competition

Creo is in the happy position now of having the entire market to itself, not counting the older technologies that it will have to replace. Those older technologies, says Spencer, are Creo's only competition right now. "No one else is making optical tape," says Spencer. "Our competition is, and always has been, magnetic tape. We're after the archival market with optical tape — it's noneraseable. These are people who want data to last a long time." Working against Creo is the fact that optical tape is more expensive, so people may compromise on technology to save money.

Free trade with the United States, says Spencer, has helped them to become more competitive. For one thing, it provided them with a customer base that is massive compared to Canada; but mainly, as Spencer puts it, "We don't like customs and duties and so on."

Management Strategy

Spencer thinks having a written business plan is important, but he never expects to stick to it too closely. "We make a new one every year," he says. "A lot of the details change, we make more revenue here, less there — you can't accomplish everything in a business plan every year."

Creo's greatest strength is its people, says Spencer, and he wouldn't be the first to say so. In the company's first decade of existence, not one employee has ever quit. "Nobody really leaves here unless we ask them to," agrees Spencer. "At least, nobody's ever quit to go to another company. One guy once quit to go open a bed-and-breakfast, but that's a whole different thing. No one has ever quit just to go across the street to a different company."

> People are usually the greatest asset of a technology firm.

In a more traditional sense, the company has always drawn much of its strength from superior technology. As the company evolves, however, Spencer sees that emphasis changing. Instead of aggressively going after the market, the company will change its tactics and become slightly more defensive, backing its sales up with topnotch service. "We expect the emphasis to shift," says Spencer. "We want to maintain that strength, but we want to have an advantage in service. We want to be very responsive to the customer. We want to get back to them right away: if a machine's down, you're over there fixing it. We've done quite well in that area — we just have to make sure we keep it up."

When it comes to management strategy, the secret to Spencer's success is a bit of a paradox. He acknowledges that his long years as a seasoned manager have certainly played a part in Creo's success; but on the other hand, many of the techniques he uses now are based on ideas he never had the chance to try out before.

"Previous management experience has definitely helped," he says. "Without it, we would have made a lot more mistakes. I'm not saying we couldn't have done it, but we definitely would have made a lot more mistakes."

Conversely, some of his theories are quite novel, and focus on how best to treat innovative people in an innovative field without squashing their creativity. "We have a flat organization, without a lot of managers," he says. "We try and train people to manage themselves. If you do that, you can grow more rapidly than if you have a big structure."

Spencer once taught a course on how to manage innovative people. "That's why we have a lot of the policies we have — open hours, no dress code, stuff like that," he says. "There's also a tolerance to failure. If you're managing creative people, there are going to be some ideas that don't work out, and if you land too hard on those people then they stop coming up with ideas. I worked at one company once where if somebody went and showed some initiative and did something brilliant that really worked out, they got rewarded — but if it failed they got their head chopped off."

In that kind of atmosphere, says Spencer, people will eventually stop taking risks. "So you have to be very careful. When an employee here does a dumb thing, you don't come down on them too hard. You still want them to try new things."

But by far the most common mistake of people launching a new high-tech enterprise, says Spencer, isn't mismanaging the people, but misjudging the market. "I would say the thing none of us pay enough attention to is understanding the market," he says. "You can't know too much about your market."

Do people get so enthused about their technology that they lose sight of whether there's actually a market for it?

"Absolutely," says Spencer. "And we've done that, too. Not so much whether there's a market, but what that market really wants. Sometimes we don't listen quite closely enough, don't call quite enough people."

To have the best people and the top technology converge at your company, says Spencer, means maintaining a good profile. "We made the decision very early that I would spend some of my time keeping the profile up in order that good people would come to us," he says.

That strategy has worked for Creo so far — to the extent that the company is always surprised on those rare occasions when it finds it has to place ads to fill positions. Most candidates are attracted to the company by word of mouth; the company relies heavily on networking to find the top people.

> Many technology entrepreneurs believe that their "mouse trap" will sell itself and do not appreciate the merits of understanding the market.

9

DY 4 Systems Inc.: Four Dynamic Engineers in Search of a Product

This case study is the story of four aggressive engineers who set up a company that floundered for a long time because of a lack of focus. Focus came through a do-or-die crisis. DY 4 survived by choosing specific market niches to become a successful company now listed on the stock market.

Swimming in money, with several months to decide which of five ways to spend it: this is the enviable position in which managers at DY 4 Systems Inc. found themselves at the end of the summer of 1993, shortly after the company went public.

The young Ottawa-based high-tech company designs and produces components that it sells to system integrators, who turn them into products useful for military, defence and aerospace purposes. Privately owned for 14 years, DY 4 began trading on the Toronto Stock Exchange in April 1993. It has grown from a four-man consulting firm to a $35 million company with 185 employees, and is now in the happy situation of being too busy taking care of business to spend much time contemplating the many potential avenues for expansion. "There are lots of opportunities for growth," says Garry Dool, the company's pleased co-founder and marketing vice-president. "An expanded sales force, acquisitions, research and development, or maybe a combination ... We're just exploring our options."

In late 1993, Dool's company was sitting pretty on some $25 million, generated when the company went public and sold just over two million shares at $12 each in its initial public offering. That makes the company a success story — in retrospect. Within four months, shares were trading at almost twice the initial amount, hovering around $24. But the company didn't get where it is today without a few hard knocks that, in the end, helped shape its current direction.

DY 4 derives its name from the four engineers who figured they were dynamic enough to go into business for themselves and succeed in the dog-eat-dog, swiftly changing world of high tech. Dool had the idea for the company in 1979 when he received a call from a local headhunter wanting to interview him for an upper management sales position. He was working for Leigh Instruments in Ottawa at the time. "After making the short list, I ended up going down to California for an interview," says Dool. "I got really enthusiastic about the idea." But he was soon told they had decided not to offer him the position because he lacked sufficient sales experience. Dool, who thought that was "just wrong," called back and spoke at length with the person who had originally interviewed him and was told they would reconsider and make another decision in a few weeks.

In the interim, he started talking to people and thinking about other opportunities. He contacted Terry Black, an engineer who was then working at Can Tech, another Ottawa high-tech firm, and seriously considering starting his own business. "We got together on a Sunday to discuss the possibilities," says Dool. "On the following Friday, I found out I had the job with the other company. I had to tell them I was leaning towards starting a new company of my own, and thank you very much."

Dool recruited two other engineers — Kim Clohessy and Steve Richards — from two other Ottawa high-tech firms, Bell Northern Research and Leigh Instruments. Dool, Black, Clohessy and Richards were the founding four who helped build DY 4 into what it is today: the largest manufacturer in its market. Over the course of its 14-year existence, DY 4 has managed to carve itself out a comfortable niche in a market that continues to grow at a rate of 30 percent to 35 percent per year — and in which there are no other major North American competitors.

The company specializes in the design and manufacture of high-end VME open architecture computer systems that automate such tasks as navigating helicopters or firing submarine torpedoes. (VME stands for Versa Modula Europa, a 10-year-old industry-standard high-end open architecture computer system.) For example, one of their products analyzes information from helicopter navigation systems and night-vision cameras. It then feeds the data to helmet-mounted visual displays which, in turn, allow pilots to fly blind.

Within this market niche, DY 4 has staked out an even more specific groove: harsh environment VME equipment. These components are designed to operate in what Dool calls "harsh, dirty, oily, nasty environments" — tanks and armoured fighting vehicles, bombers and fighter craft, missiles and high-altitude reconnaissance planes, satellites, rockets and shuttlecraft.

But when DY 4 began 14 years ago, it wasn't manufacturing anything. It was a four-man consulting company.

History and Milestones

Initially, the four engineers, each of whom had about 10 years of administrative experience behind them, sold their expertise in consulting to generate money. Black initially wanted to design a colour graphics terminal, which required some fairly heavy venture capital. "It was way too ambitious," Dool says now. "We would have needed several million dollars." So, as a consulting firm, they began to network, and attracted some big contracts. Among their first products were black-and-white personal computers. Networking steered them into designing cards customized to their clients' various requirements.

Over the next three or four years they branched into systems consulting. Noticing that many of their customers had similar problems and requirements, they began casting around for a common solution. What they came up with was the STD bus — a type of open architecture computer system still in its infancy. They began to build systems. "We were doing systems engineering, and using the funds from that to develop these parts that we could then sell all over the place," says Dool. Once they had accumulated a critical mass of parts, they began to market their product; by now, marketing and manufacturing these products had become almost a full-time job. They did eventually build a graphics terminal, but it soon became a difficult sell. Competition from US industry giants such as IBM eventually swamped DY 4's founders, sending them back to designing open system architecture cards, which they eventually sold to universities and colleges in North America. These cards are computer parts based on a common industry standard, as opposed to closed architecture cards, which are not. By 1982 — in a lesson that could be called "How to Succeed When Your Market Falls Out from Under You" — DY 4, while being driven out of the market for PCs by IBM, was ready to market a new line of military products it had been developing all along. Total revenues in the company's first three years were growing, but still modest.

That began to change in 1982, when the company — now three years old and still dabbling in consulting contracts — began to perceive a need to increase the capability of the STD system, which was based on 8-bit processors. "The 16-bit was just coming out," says Dool. "We looked around and found the VME bus was just beginning to get defined." The decision to opt for VME technology was unanimous; but still, says Dool, "We could easily have chosen one of the others available and not be where we are today."

They began by marketing VME to the military in 1983 because that's where Dool's background was. That year, networking and contacts led them to their first big contract in Denmark, which would bring in several million dollars over the next few years. That contract — from the Danish Navy — was an example of consumer need dictating technology: "We got a contract for a series of new products that was actually only a figment of our imaginations at the time," says Dool. "So we used their requirements to help define what the product would look like." That's how DY 4 launched its new product line, the VME platform for military applications. It eventually opened an office in Denmark, and sales to that country continue to be strong. The end of fiscal 1983 saw revenues reach their first significant high of $4.6 million.

Along the way, building on its growing reputation, DY 4 acquired some solid investors — CBC Pension and Noranda Enterprise Ltd., both in 1983. Noranda held 34 percent of the company and CBC Pension held 31 percent. The company began to narrow its focus on the high-end market place. Several factors influenced this decision: there was less competition in that area; and by now, most of the company's experience was in "big" systems. "We had the right recipes that these companies really liked," says Dool, "and we didn't have to compete with all those garden-variety businesses selling to the broad [commercial] customer base. We were providing capabilities that other companies weren't even thinking about yet." They were targeting customers whose needs were sophisticated enough that they would both appreciate what DY 4 had to offer and be willing to pay a premium price for it. The VME products range in price from $10,000 for low-end products to $25,000 on the high end, which is approximately 20 to 50 times the average selling price of the original STD product line. There are about 200 other vendors in the VME market, but most of those are in the commercial market, where VME products have a considerably lower selling price. By the end of 1984, revenues had climbed slowly but steadily to $5 million.

The next big contract was Raytheon Canada Ltd. in 1985. "That was about a $10 million contract, and we were about a $5 million company," says Dool. The beginning of that project heralded a grim time for the company's executives, who were busy trying to chew everything they had ambitiously bitten off without choking. Still, in retrospect, the company calls the deal a success: it made money, and Dool now looks back on it as a learning experience that both strengthened the company and represented a necessary step towards narrowing their focus. "We were still at that stage when any order is cash ... When you're growing, every order is important," says Dool. "Only when you have more cash flow can you start to focus on what it is you really want to do."

After the Raytheon contract, DY 4 began to build up external sales forces, including one in Europe, always relying on its reputation. By 1986, sales had reached $9 million. "Every contract makes the next one easier," notes Dool, who admits that if he were to do one thing differently, he would have started out with more money in the first place, instead of chasing down any and every contract that came along in the beginning.

The next major milestone was Rockwell Systems Australia Pty. in 1987, a contract worth between $16 and $20 million. DY 4 eventually won an award from Rockwell for its performance on that contract, in which it supplied data processing hardware for the combat system for the Royal Australian Navy's submarine construction program. Due in large part to this contract, 1987 sales rose to $17.5 million and spiralled to $24 million in 1988. DY 4 seemed bound for success.

But this brief moment of glory was closely followed by a moment of near-disaster that seemed to stretch on much longer. In 1989 DY 4 became heavily involved in a $10 million contract with the Indonesian government to manage a telecommunications project that was cancelled abruptly after most of DY 4's fortunes had been invested in it. In the end, DY 4 lost a great deal of money — and lost Dool as well, after a shake-up with the board of directors that followed close on the heels of the disaster. The company was plucked from the jaws of bankruptcy by its two major shareholders, Noranda and CBC Pension, each of whom kicked in $1 million.

But the money came with a condition attached: Get the company back in the black and let Danny Osadca (then vice-president of finance and chief financial officer) run it. Dool left amicably and was persuaded to return to the company in March 1993 after a three-year hiatus as general manager of Gallium Software Inc., another Ottawa-based start-up firm. Osadca, who became president, chairman and CEO, revamped the company by laying off 75 employees and refocusing the company's market niche on harsh-environment VME products. Dool now chalks the Indonesia catastrophe up to the company's inexperience in doing business with Third World countries. It taught them the importance of being relentlessly selective of their market. "After that we went exclusively after defence and aerospace companies," says Dool. "And that's what we're doing today."

> Major investors can step in to protect their investment. They often insist on professional management.

As DY 4 grew up, each of Dool's three original partners eventually left to pursue other career paths. By 1993, Richards, who left in 1985, was working with Fulcrum, another Ottawa-based high-tech firm. Black, who left in 1988 to get back into consulting, moved with his family to New Zealand in 1992. And Clohessy, who left DY 4 in 1992, is based in Phoenix, Arizona, as a consultant for the Ottawa-based Object Technologies.

Currently, DY 4's market is confined to defence and aerospace. It boasted revenues of $26.8 million in 1992, for a net income of $3.1 million. About 95 percent of its business is international; its main geographic markets are the United States, Europe and Australia. The company recently ventured into Japan, Taiwan and South Korea, and its plans for the future include forays into South America, India and Pakistan. Managers at DY 4 believe strongly in the importance of maintaining healthy international ties. "Basically, anyone who starts a company in Canada and focuses directly on the Canadian market is doomed to failure," says Dool. Because of Canada's comparatively small population, the industrial sector is just too small to offer much lucrative support. "We always planned to be international, and we are," says Dool, "to the extent, perhaps, that we've ignored the Canadian market more than we should have. But we're not sorry for it."

> A do-or-die crisis can help focus a firm on a particular area of activity.

They're still being choosy, opting to do business with only those countries they hope will be easiest to penetrate and ready to receive their products. They look for sophisticated customers because they're selling sophisticated equipment. As Dool puts it, "There's no use trying to sell them something they can't use."

But a small industry base is not the only reason for the global thrust of DY 4. Another obvious factor is the nature of their business: worldwide, military spending is in a free fall. With the exception of some areas in the Far East, most countries are not investing heavily in state-of-the-art military technology any more. Ironically, this works to DY 4's advantage: instead of buying new helicopters, for instance, a country spending less on defence can simply buy new software — the kind supplied by DY 4 — to upgrade the existing helicopters. "Say you wanted a better sound system in your car," explains Dool. "If you didn't have much money to spend, you would buy a new stereo system and install it in your old car, as opposed to buying a whole new car with a better stereo system." As defence budgets shrink, cost-effective retrofits are expected to dominate new contracts for at least the next several years.

Setbacks and Mistakes

DY 4 learned the perils of catering desperately to a broad, fuzzily defined market. In its early days, DY 4 dabbled in graphics terminals, PCs and a radar monitoring system; it designed a firebox control monitor to detect fires in the break boxes of trains in transit; exploiting STD bus technology, it also designed an electronic recorder that was never fully launched. The company solicited contracts from government and private industry, and overextended itself trying to serve each individual customer. Part of the reason for the disastrous 1989 was that DY 4 watered down its sales capabilities by not adhering tenaciously enough to its market niche, and catering instead to too large a variety of markets: the company found itself close to bankruptcy when revenues dropped to $15.8 million from a high of $24 million in the previous year, causing the company to suffer a net loss of $13.2 million. The lesson learned: Focus on a market niche with global potential.

Another lesson learned: Avoid overextension. At one point in its history, DY 4 became so successful with its STD bus that managers found the company taking on the job of system integrator as well as board designer — an example of rapid vertical integration with insufficient resources to support expanding efforts. Trying to be all things to all people resulted in the company attempting to serve too many individual customers. This in turn precipitated the decision to supply to system integrators instead, who would be able to serve the same client base far more effectively. That decision, combined with

DY 4's consequential ability to provide a faster time-to-market than its competition, ultimately resulted in the company being able to sell more systems to fewer direct clients, moving it into the top five STD bus manufacturers.

Risks and Choices

Having emerged triumphant from the wreckage of 1989 and all the earlier confusion, the company now looks as though it is set to coast to success. However, this doesn't mean its risk-taking days are over — and that sort of complacent attitude is perhaps the biggest danger. The company has identified a host of possible risks, but there are three that stand out as the biggest potential pitfalls.

The risk that looms largest is DY 4's restricted customer base. Only 10 customers accounted for 79 percent of fiscal 1992's revenues, with one customer — Texas Instruments Inc. — accounting for 29 percent of sales. The remaining 21 percent of sales was distributed among approximately 71 active accounts. This situation is not expected to change in the foreseeable future. The danger is that if any of these large customers were to cancel or delay orders, sales and income could be rapidly and adversely affected.

In addition, the high-tech market place is ruthlessly competitive: technology advancing at such a rapid pace requires the company continually to sink significant funds into research and development in order to introduce new products or enhance existing ones on a regular basis.

Because of this, the company must constantly rely on its ability to attract and maintain highly skilled, qualified personnel — and there is always the risk that these people will be unavailable or will take their talents to larger, more lucrative companies.

The Outlook for the Future

Although Dool decided to bide his time and wait a few months before dedicating the proceeds from going public to a specific area, it's clear that he will invest with the intention of increasing the company's market share in an existing niche. Currently, DY 4 enjoys about 28 percent of the harsh environment VME market, estimated in 1992 to be about 10 percent of the overall military VME market of US$107 million worldwide. That figure is expected to jump to $243 million by 1996. Dool would like to see DY 4's market share increase to around 50 percent, which he thinks is a reasonable objective. The company is projecting a 1993 net income of $4.6 million on revenues of $35 million. It now ranks among the top five in the total VME market, which is dominated by their main competitor, Radstone Teck of the UK, and Chicago-based Motorola Inc.

DY 4 divides its military market into three smaller niches: air force, army and navy, each of whose requirements are similar but different. The army needs the most rugged equipment; submarines are more benign; and ships are the easiest to accommodate, because the parts usually operate in an air-conditioned environment. Until recently, DY 4 focused on the army — whose planes represent the most demanding environment — and gained major penetration. Future plans include expansion within the navy sector. "We want to sell to the same general market (defence and aerospace), but we're trying to sell to a wider base of customers within that market," says Dool.

> As a rule of thumb, it takes $1 of working capital to support $1 of sales in technology-intensive firms. Without this level of working capital, firms meet a formidable barrier in climbing the S curve.

Going public will also help, because customers are more likely to buy from someone who is financially stable. "They want continuity, they want to know the product will still be available in five years," says Dool. Going public, Dool figures, will be the catalyst for the firm's first really visible, major growth spurt. The decision to go public was made, says Dool, because Noranda, which had supported DY 4 from the beginning, withdrew its support when it suffered hard times recently. "We needed liquidity, and we had two choices: find a buyer or go public," says Dool. "The market was looking good."

The Competition

Radstone and DY 4 fight each other for almost every major contract — and the company wouldn't have it any other way. "A competitive environment is healthy for us," says Dool. "If we were the only ones out there, we'd get too comfortable. The customer would feel locked in. They would perceive a monopoly, and they'd be far less willing to pay a premium price for our product. It keeps us on our toes," he says. Their American customer base, for instance, always ensures that DY 4 and Radstone compete for contracts. "That way, they know they'll get a competitive price," says Dool. What's odd — and fortunate — is that DY 4 has no major competitor in the United States, and none at all in Canada. "Our principal competitor was, still is, and will continue to be Radstone in the UK," says Dool. "We always know who we're going to be up against."

Ironically, DY 4 has also competed with the system integrators to whom it sells. However, this situation has ceased to represent much of a threat, because system integrators can no longer afford the luxury of designing hardware now available as off-the-shelf product from companies like DY 4. System integrators, most of whom now opt for out-sourcing, therefore represent one of DY 4's largest opportunities for increasing its business base — a trend similar to that which has occurred in the automotive industry over the last 20 years. "These companies [system integrators] have all the capability to do what we do; they're doing value added to product when they could have designed the product if they'd wanted to," says Dool. "The reason we win is they don't have the money to build enough to pay for the investment, and it would take them too long. Their delivery schedule doesn't allow them to spend 12 months developing a card, and then put it all together and deliver. They don't have the budgets any more, they don't have the people; it's not cost effective. So we see our market expanding quite rapidly in that area." There are about 200 other VME vendors, but most of those are in the commercial market; DY 4 is number one in the military aerospace market.

Research and Development

DY 4 puts about 10 percent of its earnings back into research and development. But the firm owes much of its success to the Defence Industry Productivity Program. "We've leaned heavily on DIPP," Dool says. "Having that money has really allowed us to grow." In earlier years, DY 4 also received funding from the Industrial Research Assistance Program.

Investment in R&D can allow the company to grow almost exponentially, and this is what is happening to DY 4. As Dool points out, "Once you have a certain critical mass of products, you start making money, climbing that S curve" — and then you have even more money for research and development. The company can really take off at that point, which is why the resulting growth phase is so steep and rapid. According to Dool, DY 4 has just hit that phase.

About 45 percent of DY 4's staff of 185 is currently involved in research and development. They focus on key new areas of digital signal processing and array processing. Their

Because of the high levels of R&D needed in technology-intensive sectors, firms rely on government financial assistance in the early stages of their development.

goal, as stated in the company's initial public offering, is to design leading-edge, market-driven computer products based on a common industry standard. The company's development strategy in general is to incorporate the latest enabling technologies and apply them across the full line, from commercial products through products customized for harsh environs to a full conduction-cooled product.

The company also has a strategy for specialized products: technology transfer from commercial vendors. This reduces time-to-market, development costs and support costs, and allows DY 4 to maximize its research and development expenditures while maintaining a breadth of products that distinguishes it from its competitors.

Management Strategy

DY 4's business strategy focuses on three areas: products, geographic considerations, and trends in the market place. The company has different strategies for hardware and software. In hardware, the main concern is to accommodate rapid technology turnover, to ensure system architecture flexibility and price/performance improvement. The software strategy emphasizes the provision of a full complement of software building blocks in order to provide system integrators with fully integrated hardware and software system solutions.

DY 4's geographic strategy is to expand its present sales organizations in North America and Europe along with fostering a strong network of rapidly growing markets in the Pacific Rim and Southeast Asia. The company is currently active in the United States, the United Kingdom, Germany, France, Italy, Spain, Sweden, Norway, the Netherlands, Australia, Japan, Korea, Taiwan and Canada.

Because technology changes so rapidly, DY 4 ventures require particular attention to market forecasting and product migration strategies: the initial product must be clearly defined and aggressively marketed, and — as the company discovered back in the days of graphics terminals and the STD bus — should ideally lead to other products. While keeping one eye on its innovation chain and another on opportunities, the things DY 4 watches carefully are the usual suspects: products and services, market, revenue projections and cash flow. "We're just at the bottom of the really big S curve," says Dool, "but we've got to watch a whole bunch of little S curves all going on at once, too."

According to Dool, DY 4's most important ingredient in its success has been, without a doubt, its choice of employees. "Hire the right people," he says. His company looks for dedicated, energetic, ambitious people with the right attitude — willing to put in long hours and work weekends if necessary. He says he was lucky to have many such people in the beginning, when it was especially important. "It's not uncommon for us to fly someone to Timbuktu to help out a customer. And it doesn't matter if the problem isn't with our product. The bottom line is, get the customer up and running." This approach helped win the company a prestigious award for excellence from Texas Instruments Inc. For instance, a handful of DY 4 employees once spent Easter weekend responding to an urgent, last-minute problem after General Dynamics, a contractor on a Texas Instruments M1-tank upgrade project, discovered a design problem that wasn't DY 4's fault. DY 4's mission statement is "Customer First, Quality Always."

Dool also recommends trying out your own product to see what flaws might bother the customer. "If you were manufacturing pens, but never wrote with them, you might pick one up one day and discover it's actually a real dog — it slips out of your hand, the ink runs out. Quality assurance is a big part of our business."

He also advises would-be entrepreneurs to hire people with more experience than they have, who can act as mentors. And finally, he stresses the importance of cash flow in the early days: if DY 4 had had more money when it began, its market would have been more focused earlier on. "We did a lot of bootstrapping," says Dool. "You know, do this to get money to do that. It caused us distractions along the way."

DY 4's major strengths include a solid, balanced management team with plenty of experience, combined with planned attrition and continual turnover. "We fight bureaucracy," says Dool. "We need it, but we also fight it. You have to have the right balance."

Dool believes going public is the catalyst that will set DY 4 off on a rapid growth spurt that will have to be managed carefully. The best way to do that, he says, will be to keep an entrepreneurial edge even while the company expands and management grows more professional. "You don't want to stagnate," he warns. It's not uncommon for growing companies to experience a bureaucracy that spirals out of control and decision-making processes that slow to a crawl; this is a situation to be avoided at all costs. "We have a lot of entrepreneurial spirit left in us," says Dool. "We have to keep striving to be innovative, creative enough to beat the competition and be first." Company president Danny Osadca's philosophy is well known to the firm's executives: he attributes DY 4's success to innovation, aggressiveness and a willingness to take chances. He doesn't see that approach changing regardless of how big the company gets.

And finally, a word on attitude: Dool says he has always believed a good offensive is the best defence. DY 4 is still too young to have grown complacent in its niche; it's still trying to carve out a deeper market share than what it has.

Dool sums up the spirit and essence of getting ahead in high tech: "We're always improving our product, improving our ability to turn out new products faster. That way you can make your competition work harder. We can win more market share by delivering a product three months ahead of someone else whose system isn't as streamlined."

> To be successful, firms have to be aggressive in marketing and product development. Moving from one S curve to the next has to be part of corporate culture.

10

Corel Corp.: Entrepreneurship, Experience and Deep Pockets

Michael Cowpland launched Corel with $7 million that he got from the sale of his previous venture, Mitel. By focusing on a leading-edge product early on, and with aggressive international marketing and discipline, the firm became a $100 million company after only seven years.

Building on essentially one type of software product that began as a sideline to the company's "real" business, Corel has grown in only seven years into a $100 million company that employs more than 300 people and is a world leader in manufacturing graphics packages for PCs.

Although it started out designing a variety of products, including a hardware line and optical disks (CD-ROMs), Corel's key products now are software packages for graphics on PCs. These programs are used to design a wide variety of items — such as business cards, flyers and restaurant menus — and to make charts, edit photographs and prepare slide shows. Essentially, the various versions of CorelDraw allow even the most computer-illiterate, unartistic people to play with words and full-colour images on a computer monitor, and then print the finished design out using a laser printer. Words and letters can be flattened, expanded, lengthened or braided around the page in wave patterns — almost anything imaginable is possible, including any picture a user is inspired to draw with a mouse. In late 1993, Corel introduced PhotoCDs. Operating on a CD-ROM disk drive, each CD contains 100 full-colour images that can be retrieved and used in illustrations. Corel's products continue to sweep up annual industry awards in magazines around the world for the best graphics package; the company has won more than 90 awards for best product.

Corel is Michael Cowpland's baby, a project he launched in late 1985, shortly after his glory days as co-founder of Mitel Corporation came to an unceremonious end. He set up Corel with $7 million of the proceeds he earned from the sale of Mitel. Cowpland — who launched himself on the Ottawa high-tech scene after graduating from Carleton University with a PhD in engineering — is often referred to as Canada's answer to Macintosh co-founder Steve Jobs. He has become a mythical figure in Ottawa, where even people who aren't involved in the high-tech industry will tell you he's a maverick businessman whose every touch brings forth gold. The media have characterized him as ruthless, brilliant, powerful, arrogant, shrewd and any number of other terms, not always flattering.

> Not many entrepreneurs have large amounts of money in hand to launch their ventures.

After spending a few years at Microsystems International Ltd., a division of Northern Telecom that eventually folded, Cowpland co-founded Mitel, a telecommunications firm, with partner Terrence Matthews. In its heyday, Mitel was often referred to as a high-tech wunderkind. The two built the company from scratch into a $350 million, 5,000-employee empire before it spiralled out of control and then almost fizzled out of business. British Telecom finally stepped in and saved the company, buying it for $320 million early in 1985; Cowpland stepped down as chairman, although he remained a board member. Matthews went on to found Newbridge, which got off to a bumpy start but began making money hand over fist in the early 1990s (see Chapter 11).

Cowpland couldn't stay away from the adrenaline rush of entrepreneurship for long. He kept a low profile for just less than a year after Mitel, then he put some of the money earned from the sale of Mitel into his new venture, and resurfaced as the founder of Corel Systems Corporation in 1986. He focused initially on laser printers, because it looked like that market was just opening up at the time. However, he hadn't counted on his potential customers — 30,000 Micom AES word-processor users — disliking their systems so much that they would refuse to sink any further funds into upgrading them. By the time he was ready to sell his printers, nobody wanted to buy them.

So he forged ahead with a new twist on laser printers: "dumb" ones that would be compatible with any PC. Shortly thereafter, another obstacle blocked his path: these printers could not operate on a network — and networking was becoming increasingly crucial, because PC use in businesses was growing exponentially.

Cowpland's answer was desktop publishing systems — a brand new field, at the time. He planned to build topnotch systems and service them. He marketed Corel's new systems through a dealer network and monitored customer feedback closely. He also happened to own 20 percent of a Toronto-based company that had been developing desktop publishing technology for several years, and he continued to spend money on development costs.

By the end of 1987, Corel's hardware sales were about $5 million — roughly half of what Cowpland had originally predicted. Software sales had unexpectedly accounted for a further $1 million. Meanwhile, engineers were still masterminding the software that would make Corel an industry standard. By the end of 1989, they had come up with a wildly successful graphics program that was selling briskly in 17 different countries. Sales of CorelDraw jumped from $7 million in 1988 to $17 million in 1989. By halfway through 1989, Corel's software sales were growing at a rate of 150 percent per year.

Corel's greatest strength, says Cowpland, has been its ability to focus on providing a good product in a quick turnaround time, and then upgrading it on an ongoing basis. "Also, it's been the ability to spot where the industry was going and get there first," he says. "For instance, in the case of graphics, we thought all-in-one graphics would be a good idea. So we did that with CorelDraw 3.0, and it's now turned out to be a great idea. So we did it even more with CorelDraw 4.0. As a result, we've leapt ahead to number one in the world.

"Same with CD-ROMS," he continues. "We spotted the trend towards CD-ROM before other people did, put all our products on CD-ROM, and now we're number one in shipment of disks. We've seen the same thing happening with PhotoCD. We saw this was the best way to do it, leapt on the opportunity and now we're number one in shipment of photos on disk. We now have 200 titles, and we're producing one new title per day."

History and Milestones

"The first few years [at Corel] were pretty tough going because we were feeling our way," Cowpland says. "It was a brand new business and we were finding our way around it. Eventually we got into a systems oriented business, where we were putting together equipment from other vendors but adding special utility software to make it better. We'd build the systems around Ventura or WordPerfect and give people desktop publishing systems including computers and lasers. We started making our own lasers."

However, it soon became apparent that the systems business was going to be a slow starter. "That was change one," he says, about designing their own lasers, "and then change two was, we found the systems business was fairly tough as well, because it was hard to get any ongoing value added," says Cowpland. "Each sale was a sale, but then didn't particularly lead to another sale, it was just one sale at a time. And then we had software modules that we were bundling with the systems that made ours unique, and we decided to sell those separately as utilities — $50 or $100, compared with a $25,000 system. So it seemed like that would be puny by comparison. But we rapidly discovered that because of the huge world market, software was very good business, so we decided to concentrate more on the software than on the systems integration. We started the CorelDraw project, and after 18 months of development, that emerged and was immediately a winner."

> Systems integration, like consulting, is a relatively common approach to starting a firm.

It took three engineers about a year and a half to develop version 1.0 of CorelDraw, which was finally introduced in January 1989. In that first year, Cowpland was counting on selling 5,000 units. Instead, he sold 50,000, and a small marketing team managed to establish CorelDraw as the standard in both Britain and Germany. In November 1989, Corel went public, trading on the Toronto Stock Exchange and raising $20 million; shares began selling for $7.50. All three of its divisions — systems integration for desktop publishing, local area networks and presentation, and optical disks products to design, manufacture and sell interface software — were still profitable, but CorelDraw was clearly taking a big lead, earning the company profit margins of about 80 percent. ("With systems," says Cowpland, "you're lucky to get 30 percent.") By the end of 1989, CorelDraw was selling in 19 different countries. Corel had set up alliances with several large distributors in the United States and Europe.

> A winning product is often not the one that was originally in the entrepreneur's mind's eye.

In the following year, 1990, the company followed its first product up with the new and improved CorelDraw 2.0. "If you look at just about any innovation," Cowpland says, "it's the second or third one that works. The Apple II computer was the winner, not the Apple I. The Mac is really a smaller version of Apple's Lisa computer, which was a failure." The secret, he says, is to "find a field you like and have an entry vehicle — realizing that you'll more than likely be changing directions."

By 1991, CorelDraw software was accounting for almost all of the company's $52 million in sales — an increase of about 78 percent from 1990. Corel released CorelSCSI (Small Computer System Interface) that year, allowing users to connect as many as seven peripherals to a single host adapter. (In 1992, a software-only version was launched, and has since become the most widely used SCSI software solution on the market.)

Corel had a banner year in 1992. The company gained more than a 70 percent market share for its graphics packages in both Britain and Germany, and made significant progress in Poland, Hungary and Czechoslovakia. Sales grew by almost 80 percent. The company was cited in *Profit: The Magazine for Canadian Entrepreneurs* as one of Canada's four fastest growing companies. CorelDraw 3.0 was launched early that year, along with the new version of CorelSCSI. That same year, Corel discontinued its hardware sales and decided to focus even more closely on software development. Cowpland decided to drop the word "Systems" from the company's name, switching simply to Corel Corporation, to signify the new, strategic direction the company was taking. Its new goal was — and still is — to make CorelDraw the graphics standard around the world. By then, more than 300,000 Corel units had been sold worldwide; by the end of 1993, the number had grown to 550,000.

CorelDraw 4.0 replaced version 3.0 early in 1993. Some of its improved features included animation, seven modules, optical character recognition, 400 more fonts, new drawing tools, and more consistent interfacing. To avoid confusion caused by the rapid rate at which new versions of CorelDraw replace previous ones, Cowpland kept the 3.0 version on the market, cutting the price back to $199.

A well-focused product-migration strategy brings results.

Corel also came out with support for Kodak's PhotoCD technology, with both CorelSCSI and CorelDraw, that year. The software allowed desktop publishers to produce photograph-quality images by combining the quality of film with the flexibility of a digital format. Another development was CorelShow Runtime Player for multimedia presentations. Corel also introduced 24-hour support to registered users of CorelDraw, Monday to Saturday.

Corel has never received financial assistance from the government. Cowpland has lobbied unsuccessfully for research and development funds, leading him to conclude that it's difficult for a high-tech company to establish itself in Canada — and to stay there — when there are so many more incentives to move to the United States.

Corel had about 330 employees by the end of 1993. "We're trying hard to keep it down, because we like a fast-moving company, and the more people you get, the more you get bogged down by bureaucracy," says Cowpland. "We're all in one building here too, except for a small group in Dublin, Ireland, to cover Europe. Employees are growing at a much slower rate than sales." In the quarter just finished, Corel was number one in the industry worldwide in product-per-person.

As for research and development, Cowpland says Corel aims to have a 12-month horizon for most of its spending. "In other words we like to be able to choose an objective and make it happen within a year," he says. "We do some background work on things that aren't necessarily products, but might be. For instance, we just developed an in-house database that's running the whole company and that may or may not become a product." The company currently spends about 7 percent of its revenue on R&D.

Small businesses make up the largest part of Corel's market, although its products are also used by individuals and by large multinationals. The company has a network of some 80 distributors in 40 countries, and derives about 95 percent of its revenues from exports. About half of its market is in the United States, another 40 percent in Europe, 5 percent in Canada and the remaining 5 percent elsewhere in the world.

> Marketing expenditures become very significant in the steep part of the S curve.

Corel spends an average of 20 percent of its revenues on marketing and promotion. This is largely due to Cowpland's personal strategy: ideally, a company should find a market niche that is still largely unexplored, develop the best possible product and then amplify the product's potential with aggressive marketing. "A company can easily treble or quadruple the strength of a good product with marketing," Cowpland says.

Another reason for the company's runaway success has been its ability to provide leading-edge software at a competitive price. Its main competitors — like Micrografx — haven't been able to keep up.

Setbacks and Mistakes

The initial investment in getting to commercial production was more expensive than Cowpland expected, he says. "It took about $6 or $7 million in working capital, which was about 10 times more than I expected, to actually get to the point where there began to be positive cash flow." This is why, he points out, you have to be cautious and manage your capital carefully to start up a business and make it survive. "There are a lot of surprises," he says. "Fortunately, I had deep enough pockets that I could do it on my own. But most people couldn't have — and they would have been out of luck, probably have had to sell the company, or get diluted down to nothing by venture capitalists."

> A company is usually more time-consuming and expensive to start up than the entrepreneur expects.

Corel was originally involved with both hardware and software products. However, it wasn't long before it became obvious that the road to success would not be paved with hardware's declining profit margins and easy-to-manufacture clones. So Cowpland made an early decision to narrow the focus and concentrate on software, for which copyrights could be obtained to protect the product for at least a few years.

In its earliest years, Corel learned a few valuable lessons about the dangers of expanding too rapidly when it introduced its products to the European market. Lack of foresight caused them to neglect the potential damage to Corel's reputation and bank account that could have been caused by not taking the time or money to translate CorelDraw into more than one language: they launched CorelDraw into an international, culturally diverse market when the product was still available only in English. The result was much slower sales than expected, along with some temporary image problems. Corel retreated to do some damage repair; by summer 1993, CorelDraw was available in 12 different languages, including French, Italian, German and Portuguese. Since then, the company has added an additional 10 languages.

> Globalization of the world economy means that Canadian firms have to be more sensitive to different cultures and languages. This sensitivity is also good for business.

"We find that in typical countries, you can sell English because there's always somebody who speaks English and will buy it," says Cowpland. "But if you take the trouble to translate, it normally doubles or trebles the sales. So we've translated it into 22 languages."

Because of the way Corel toyed with hardware for the first few years, only to drop it entirely later on, the casual observer would think Corel made a few unfortunate wrong turns in those start-up years. According to Cowpland, the casual observer would be wrong. He fully expected the first product or two to flop — or at least, not to be the successes that would put the company on the map. Both Mitel and Corel, Cowpland points out, started off making products that were very different from those that finally became big winners.

Risks and Choices

Corel relies quite extensively on only a few products that have short life cycles. It would seem as though the consequences could be devastating in the event of an unforeseen disaster, such as declining sales, a better software package made by someone else, or a saturated market. Some analysts think this is a good thing: it means Corel has focused on one revenue stream and plans to continually expand that clearly identified market with new products. Others caution that the market is full of up-and-coming competitors, and if Corel fails to produce exactly the right new package, it will have nothing else to fall back on.

Cowpland isn't worried. His theory is that it's smarter to keep a narrow focus — to corner and expand the market in a specific area, and then continue to follow those original products up with improved, leading-edge versions. To diversify at the expense of concentration would make that impossible to do. So far, the formula seems to be working. "It's a huge niche, and also we've continued to broaden it with other products — Corel Ventura, and the PhotoCD titles, for instance — so we know we've got quite a wide product range, although CorelDraw continues to be the star project," says Cowpland. "We've got two versions of Draw on the market right now [3.0 and 4.0] and CorelDraw 5.0 is coming out [in 1994], so we'll have three on the market at once." CorelDraw 5.0 was released in May of 1994. With this version, Corel becomes a bigger player in the desktop publishing market by combining the graphics capabilities of CorelDraw with the publishing capabilities of Ventura. CorelDraw 5.0 integrates the two in an all-in-one package with some 22,000 clip-art symbols and 120 PhotoCD images.

The Outlook for the Future

Cowpland intends to continue to market Corel's software aggressively to small businesses, expanding his share of that niche. The company is still in the process of getting rid of its hardware inventories. Other plans for the future include new strategic alliances, of which the company has already made about a dozen, most of them for marketing purposes. Macintosh-compatible products are also being planned. "A Macintosh product will be coming out late next year [August 1994] and we will have a product for Macintosh probably within about three months, which will be basically clip art plus a viewer, so we'll get the ball rolling," says Cowpland. "We've also got the PhotoCDs on the Mac, of course."

> Adapting a successful product to another established market is a relatively surprise-free strategy.

There's also the possibility that it won't be the big seller Corel anticipates, because Adobe Systems Inc. already dominates the Macintosh market. Another innovation is Corel PhotoCD. The company introduced 40 titles late in 1993, and expected to be shipping about 200 by the end of that fiscal year.

Cowpland also has plans for further international expansion, particularly within Germany, where Corel made a big splash and a lasting impression early in its history. To strengthen its presence in the rapidly expanding European market overall, Corel incorporated a wholly owned subsidiary company in Ireland. Also, it acquired Ventura Software Inc. and all its related technologies, including Ventura Publisher and Ventura DataBase Publisher.

Cowpland expects CorelDraw to continue to generate more than 75 percent of Corel's sales in 1994. As for market share, "We'd like to get to about 60 or 70 percent," he says. "Right now we have about 50 percent." CorelSCSI should do well because despite its relatively small market, there are virtually no competitors.

There are no plans in the works now for acquisitions, says Cowpland. "We don't really plan for them, but if they come up we may do them. And we don't buy companies — we buy technologies."

Cowpland identifies the company's biggest challenge as staying streamlined and efficient as it continues to expand. "The danger is that as we get more complexity and more products, and more deals cooking, we could get less focused and less effective in carrying them off," he says. "We have to make sure we don't lose our sharpness."

The Competition

Some of Corel's main competitors have historically been Software Publishing Inc. of Santa Clara, California (whose parent company derives 80 percent of its $170 million annual revenue from a graphics package called Harvard Graphics); Aldus Corp. of Seattle, Washington; and Adobe Systems Inc. of Mountain View, California, whose Illustrator software is popular with many Macintosh users.

Another, more aggressive competitor emerged recently: an American company called Micrografx Inc., of Texas. It's estimated that Micrografx claimed only a 7 percent share of the market, but the company's aggressive marketing tactics garnered it more and more attention from Corel's marketing watchdogs. For instance, Micrografx once ran ads claiming that Corel's graphics software was "wimpy, wimpy, wimpy."

> Head-on competition can lead to very aggressive marketing approaches.

"Why waste your time with a wimpy product when you can really flex your muscles with Designer?" asked the ad, placed in an American computer magazine in late 1991. Micrografx's Designer package competes directly with CorelDraw. Corel responded by placing ads pointing out the differences and drawing attention to Micrografx's meagre 7 percent market share; Cowpland dismissed their antics as "wishful thinking." Micrografx reported a 50 percent increase in revenues from 1991 to 1992, earning $47 million U.S.

By 1993, Micrografx was still making inroads on Corel's market. It competed mainly by dividing its graphics products into two distinct markets and selling one of them — Graphic Works — for about half the price of CorelDraw. (Its premium package, Designer, is the equivalent of CorelDraw and is sold separately for a higher price.) Corel retaliated by leaving older versions of CorelDraw available on the market at a greatly discounted price, undercutting Designer. Micrografx has responded to competition from Corel by imitating their marketing strategy. Noticing how successful the launch of CorelDraw 3.0 was in the spring of 1992, combining high-end graphics with graphics for both the untrained and the unartistic, Micrografx offered a similar package later that year. Now they sell a package called Graphics Works, a combination of low-end products. They still sell the premium Designer package separately. However, says Cowpland, that company no longer represents much of a threat. "[In 1992] our main competition was Micrografx, but now they're no longer very competitive, because they lost $2.5 million in the last quarter [ended summer 1993]," he says.

To stay ahead of the competition, Corel also plans to release new versions of its software packages every year. With CorelDraw 4.0, which included desktop publishing and animation features, it had also begun to chip away at the market previously dominated by desktop publishing companies like Quark and Aldus. "We think our next big competitor will be Aldus, because we've got Ventura and they've got Pagemaker," says Cowpland. "So that'll be the top competitor for [1994]."

Management Strategy

Cowpland has this advice about getting a new enterprise off the ground: "I always tell people if they're planning to set up a business, keep the powder dry — don't spend too much money on the first ideas, because chances are they won't work out as you expected," says Cowpland. "You have to evolve rapidly, and it's normally either the second or the third try that works."

Cowpland's whole management strategy is built upon one central tenet: Avoid bureaucracy. "We try to keep the structure as flat as possible," he says. The company only has three "layers" of staff, and Cowpland is one of them. "I have eight people reporting to me, and each of them has up to 20 or 30, and then there's the bottom level."

Cowpland wants the company to remain smaller than 1,000 employees. He says that once Mitel grew past that point, he began to get bored because his job turned into "one meeting after another" while the excitement of entrepreneurship melted into bureaucracy. Cowpland makes the time to interview every new employee — he has more than 300 now — and likes the fact that he knows at least 100 of them very well.

Flexibility is essential in technology firms. Too much structure can inhibit innovation. IBM is a prime example of a company that has suffered due to heavy bureaucracy.

"I believe it's important because it's the people who make the company," he says. "If you do meet them all, you show that the hiring standards are being kept up. For example, I normally find that engineers, because of the enormous amount of screening they take before they get here, almost always work out. On the other hand, it's different if it's somebody at the administration level. I found at one point that they were letting the standards way down. It's important to have the right people at the beginning, because before long, they're doing senior work. Then you could have a lot of problems, and you've got to move them out again. Better to correct that at the beginning. And by me taking that trouble, other managers take the trouble too, and the standards stay very high. At this point, I rarely veto anybody; it's more like just meeting them, which is fine. And I like to meet them anyway."

The employee phone directory features pictures of each staff member, listed by first name. "It's nice to know people's first names right away," says Cowpland.

Part of his well-known management technique involves accessibility. Gone are the days when he occupied an impressive corner office at Mitel. These days, his office is planted right in the middle of the Corel building — employees call it "The Pit," or "The Fishbowl" — and the door is open to anyone, any time. There are no secretaries or assistants. "I believe in being in the middle of the company as opposed to a nice big office looking out the window," he says. "I like to be at the centre of what's going on, because it's important to keep in touch."

A big part of Corel's success formula is aggressive marketing, for which no expense is spared. Corel spends more than $15 million a year, advertising in about 140 magazines around the world, and spends another $2 million translating its software into 22 languages. Cowpland believes being first into the market is crucial at any cost, and so in 1992 Corel began offering a 50 percent discount to all buyers in Eastern Europe and other areas where people couldn't afford the usual price. This move was in line with Cowpland's pet theory: Find a niche that isn't dominated by anyone, develop the best possible product and then magnify its potential with marketing. In other words, the discount was considered a worthwhile sacrifice for being first into that market. Corel has since dropped this plan. Instead, it has begun selling the CD-ROM version of CorelDraw 3.0 for only $149.

> Being first to market is the strategy of most technology firms. This requires leading-edge products and aggressive marketing.

Listening attentively and responding to customer feedback has also played a big role in some of the improvements that have made CorelDraw such a success. Corel concentrates its efforts on marketing and R&D, and subcontracts other operations, such as manufacturing and shipping. It is to this strategy that the company feels it owes its enormous profits: in 1992 it had a sales-per-employee ratio of $330,000.

Sales growth for CorelDraw, says Cowpland, is about a 10-year curve. "We're about halfway up it now," he says. "The best way we've found for managing such rapid growth is to have a completely flexible organization that accepts changes as a matter of course — nobody minds being shifted around to different job functions, and we let the best people get ahead. If people don't fit the organization, we encourage them to leave."

Corel's fortunes don't have the federal government to thank for much. Cowpland's attitude about government R&D spending is more dismissive than anything else. "We don't bother to lobby for funding because we find there's too much paperwork involved," he says. "Also, one's own targets tend to move too quickly. By the time you get people to agree to fund whatever it is you want to do, you've changed your mind on what you want to do. Then you're forced to do almost exactly what you said you were going to do just to get the money. That's not good for business. It impedes flexibility. I mean, if we get money from someone to do systems, do we have to pay back the money if we decide to do something different? You lose flexibility, and it's hardly worth the trouble.

> Government time frames are often in conflict with those of fast-moving firms.

"Better to do it more slowly if you have to," he advises. "Otherwise, you tend to lose focus of your objectives — you spend too much time getting grants instead of getting customers."

11

Newbridge Networks Corp.: From Near Disaster to Success

Newbridge is the story of Terrence Matthews, visionary and marketer par excellence. He launched his company in 1986 with $14 million of his own money, obtained from selling his interest in Mitel, a company he co-founded. The firm grew rapidly, confirming the founder's vision. However, product performance difficulties led to near disaster. Deep pockets and management changes saved the day. Newbridge has since returned to spectacular growth.

Thriving is an inadequate description of Newbridge these days. Founded in 1986, the $300 million telecommunications company has 1,500 employees, is hiring at the rate of one new employee per day and is growing at about 60 percent per annum.

Newbridge's products, which are electronic switches, are tough to describe in terms that would make any sense to someone with no background in technology, says Jim Mackie, vice-president of business development at Newbridge. "Our products are most easily described as highway interchanges to the information highway," he says. "People seem to understand that the information highway is key to tomorrow's economy, and that's where we fit in. We don't really build the superhighways; but when they become small roads, or in our case, paths — bicycle paths, trails — we handle those transitions, both upwards and downwards."

Mackie previously worked at Mitel as a "telecommunications futurist," ensuring that the company's technology was taking it where he anticipated it ought to go. Newbridge's products take small communications channels and make them into "bigger and bigger and bigger ones," says Mackie, "and we take big ones and make them into smaller and smaller and smaller ones." This is useful for a variety of businesses in areas from small operations within a company to wide metro areas and international networks for companies. Newbridge supports a number of initiatives worldwide, with telephone companies providing networks for banks or large corporations, such as the Financial Network Associates (FNA). There are a number of other networks in the planning stages and in implementation now, with the same objectives: the idea is to provide a digital connection for voice, data, image, video, and local area network traffic onto a big "highway" that's either privately operated — in which case the company rents the services from the telephone company — or publicly provided by a telephone company.

Newbridge was founded by Terrence Matthews, previously the co-founder of Mitel Corporation. Matthews, who grew up in Newbridge, Wales, and came to Ottawa in 1969, first worked at Microsystems International Ltd. for several years before quitting to start Mitel with his colleague, Michael Cowpland. He resigned as president of Mitel in October of 1985 and started up Newbridge. He's now reportedly bent on avoiding his past mistakes and outshining even Mitel, which in its glory days was considered to be the Canadian high-tech success story of the decade.

His tactics will be to pay closer attention to finances — controlling them more carefully during rapid growth — and to avoid another fiasco like the famed SX2000 one (described in Chapter 14), in which Mitel's inability to deliver its major new product as promised shook the confidence of both customers and investors, and put the company on shaky financial ground. In 1986, Mitel was bailed out by British Telecommunications PLC, which purchased 51 percent of the company for $322 million. Having lost a controlling interest in the company, both former partners began to pursue other entrepreneurial interests, which lead Cowpland to found Corel Corporation and Matthews to start up Newbridge.

Although Newbridge enjoyed great success almost from day one, it went through a rough period in 1991. That was when Matthews finally learned a few lessons about over-promising results. In 1991, buffeted by too-rapid growth, targeted by aggressive short-sellers, and troubled by software bugs in equipment already sold to crucial customers, Newbridge lost $18 million. That was when Matthews really began to avoid reporters, although rumour has it he'd been instructed by the board of directors a year before to keep a lid on the unattainable public forecasts he was prone to delivering. Although he generated enthusiasm in abundance when it was needed most, he routinely disappointed shareholders and customers. These days, he can't be contacted personally by journalists, who are firmly instructed to go through the public relations department first. No Newbridge employee can speak to the press without prior approval, either from the public relations department or from Matthews himself. Since 1991, those who have spoken without authorization have consistently refused to give their names, even to say something great about the company.

> It is sometimes difficult to escape the past, even in the fast-moving high-tech world, especially when patterns seem to repeat themselves.

The ghosts of Mitel's greatest mistakes still hover over Newbridge's reputation with a persistence that has not touched Cowpland or his post-Mitel venture, Corel. This is partly because Newbridge's products are in the same family — telecommunications — as Mitel's, whereas Cowpland's venture is exploring completely new territory. Newbridge, conversely, seems to have unwittingly done everything in its power to invite comparisons. It was launched with mostly former Mitel people. It has since hired scores of former Mitel people. By 1989, 10 of Newbridge's 12 senior managers were from Mitel, and even today that number hasn't dropped below half. It is situated directly across from Mitel's original plant, which it eventually occupied when it expanded in 1989. And more significantly, its growth pattern, fast and furious, parallels the early days at Mitel.

History and Milestones

Matthews worked for British Telecom for about a decade, earning an engineering degree in the process, before he left the UK for Canada in 1969 on a holiday. He and his wife liked Ottawa so much that they sold their return tickets and stayed. Matthews got a job in the marketing department of Microsystems International, where he met Cowpland. It wasn't long before the two were talking about starting their own company. Mitel was incorporated in 1971. By 1973, when they finally got Mitel off the ground, the idea had become reality.

Matthews was still on the Mitel board, watching his company sink further and further into debt, when he incorporated Newbridge in 1986 as a separate, private business venture. In an effort to facilitate cash flow and avoid the pressure of finding backers, he used $14 million of his own money to finance the initial research and development investment. After that, employees chipped in a further $7 million, and owned some 20 percent of existing shares by 1988, a year before the company went public.

Newbridge started off providing messaging and routing networks, differentiating them from those of competitors by being more flexible. The products were low-cost PC local area network controllers and multiplexers — black-box devices used to route data and voice simultaneously down digital communications networks. These devices allow faxes, photos, computer data, video and someone else's voice to travel down phone lines; sent all at the same time together, they arrive at the receiver's end neatly separated and ready for use. Newbridge built them to suit individual customers' standards, whereas competitors were building them one way only and leaving purchasers to modify them. The company expected to be profitable by 1989.

Some of Newbridge's first customers were Sears, Sony and MCI Telecommunications in the United States, and the federal government and Rogers Communications in Canada. Overseas, they were British Gas PLC, British Rail and the Marconi Company Ltd. The first, most important sale was to the Eastern Electricity Board in Britain.

"From the very beginning, we've had a strong view about the software that manages the network — that's done through a high-performance work station, where someone sits and gets a complete view of all the elements of the network," says Mackie, describing the evolution of Newbridge's product line. "Those elements, obviously, started out being rather simple a few years ago, but now they're much more complex, and embody technologies at the local area network. We weren't in that area before — we were a wide area networking company originally. So now we've moved our vision more into the management of devices that are in the local area network."

As a consequence, the products have also had to become much faster. "The highways have become faster and faster and our products have had to go faster to keep up with them," says Mackie. In technical terms, that means a jump from T1 or E1 links, which move at only one megabit and a half per second, to T3 links, which operate in the range of 45 megabits per second. The E1 and T1 links have been the building blocks of high-speed digital communications in the past — the past, as Mackie points out, meaning up to about three years ago.

"More recently, we've moved into a transmission scheme that has no speed limit," says Mackie. "It's called asynchronous transfer mode (ATM) — a transmission technology that treats voice, computer data and TV signals equally. The revolution here is that it's the first time there's ever been a standard that computer, telephone and television people have all agreed upon as being something they could work with. It's revolutionary," he enthuses. "ATM is where people see a great vision in services to the home and business. We have a leadership position now in the world in supplying technology like that."

> Opportunities in the telecommunications sector abound due to the convergence of computer, information and communications technologies. Canada has major strengths in this area.

By 1989, the company had grown to 1,000 employees, with $67 million in sales — four times the figure for the previous year. Newbridge won the Canada Export Award that year. Growth at Newbridge was breathtaking. Employees in every department worked massive amounts of overtime just trying to get products shipped on schedule, striving to keep up with the rising demand. Worldwide staff doubled that year, and Matthews set up a network of 35 sales offices around the world, and plants in the US, Britain, Puerto Rico and Hong Kong. The speed of the expansion left a few loose strings dangling — product glitches, unqualified employees who'd been hired on a moment's notice, and complications with international telephone regulations. Matthews had gone into the massive expansion phase with a total of close to $30 million, raised by institutional investors and staff. He came out of it desperately needing more. Newbridge had become a bottomless consumer of funds. A public offering in the fall of 1989 raised a further $46.7 million. Shares began trading at $12.45 and took off from there — within three months, they had reached almost $20.

By 1990, exports accounted for 94 percent of total sales, and Newbridge had become the third biggest telecommunications company in the world. It was doing a steadily growing business in 35 different countries, from Europe to the Middle East to Africa. Everything seemed to be cruising along at a swift, efficient pace, accelerating with every step. This rapidly spiralling growth, however, is exactly what worried investors and analysts, who had seen it all before ("it was called Mitel," said one analyst) and wondered how closely history would repeat itself.

Anticipating continued rapid growth, Matthews had installed enough offices and plants around the world early on to support $200 million in sales by 1990. However, 1989 had seen only $67 million, and 1990 only $120 million — not bad for such a young company, but not enough to justify the aggressive expansion. There were some signals, in 1990, that Newbridge's market was leveling off: the bigger companies had been responsible for most of Newbridge's early growth, but now they had their systems in place and weren't spending any more money; and there were simply too many new, similar products on the market now for the smaller companies to spend much money on any of them. The recession was also a factor. Meanwhile, the worldwide expansion that had begun a year earlier was still consuming an astonishing amount of money. Newbridge's expenses had doubled by the end of fiscal 1990.

> Rapid growth requires the production and marketing capacity to sustain that growth. If the sales do not materialize then the company is in trouble.

Expenses continued to be higher than revenues for much of 1990 and 1991, so that by 1991 Newbridge's stock was limping along at $4. Within two years, however, a turnaround had taken place, and the company was finally beginning to dominate the market it had created. The world's biggest telephone companies were all starting to pay attention. Stocks began to look a little healthier, and ultimately reached a high of about $113 by June 1993.

What magic brought about this positive turn of events?

"Some of the things that happened in that time frame were really interesting," says Mackie, referring to the period between 1991 and late 1993. "In the old days [at Mitel], our products were largely sold to private network operators, including some of the biggest companies in the world. In 1983–84, Thatcher liberalized the British telecommunications market. In the US, a judge ordered that AT&T be broken up, creating a situation that allowed private networking people to purchase communications capacity from both regional Bell operating companies and AT&T. At that same time — 1984–85, around the germ of the idea of Newbridge — there was very little equipment around, and no manufacturers, to optimize that capability." That was where — and why — Newbridge came in.

> Major changes in an established market create opportunities for new players.

However, it wasn't long before telephone companies began to realize they were losing customers because people were buying their communications capacities wholesale and running their own networks. The problem for telephone companies became one of how to get these people back. To woo them back, says Mackie, Newbridge wanted to offer former private network operators the same services they would have run on their own, but with the telephone company providing them. "Interestingly, right around the same time, the early 1990s, some of the people who had classically, traditionally run their own networks were getting very big, and their networks were getting pretty complicated," says Mackie. "So people were saying, 'Gosh, do I need this trouble, with all these people and all this equipment? It's not really my business. I don't run a telephone company.' So the phone company went back to these people and said, 'Look, we'll take care of all this for you using our products.'"

For Newbridge, the result was a changed market. Increasingly, sales were to telephone companies instead of private network operators. "The big transition is sort of a 70–30 percent shift," says Mackie, "with 70 percent of our products being sold in the early days to private networking people, whereas this year, about 60 percent are sold to telephone companies, and only 40 percent to private networks. So there's been almost a complete shift in the profile of our customers in the period of the last three years."

Although there has been some fine-tuning, Newbridge's products haven't changed much in purpose or targeted market since the company's inception. "The core product line is the same, philosophically," he says. "There haven't been any changes in the general approach to the product or how it's managed. We talk about the network being our product rather than the boxes we sell being our product. Although that sounds like marketing noise, it's true.

"What customers get with our product is the ability to create a very large network and manage it in fine, fine detail." This, he says, is partly what distinguishes Newbridge from its weaker competitors. "Many other solutions would just have you connecting a number of boxes together, and keeping your fingers crossed over whether or not it was going to work. And then, if something fails, you get out, get in your truck, drive there and try to figure out what's wrong."

Newbridge didn't receive much in the way of government funding as a start-up company, relying mostly on Matthews's considerable wealth to back it up. "We're not exactly a textbook example of how to start up a company," says Mackie. Because of Matthews's initial $14 million investment, the company didn't rely on government money in its start-up phase. But once it got rolling, it began to make use of available funds to further stimulate research and development efforts.

"We consider government funds to be a stimulus to what we would otherwise do," says Mackie. "So we're not dependent upon government funding to do any research — we're going to do it anyway. What the government funds do is accelerate the task of getting it done, and in that sense they're very beneficial. In every case, royalty streams are involved, so the taxpayer gets paid back out of the success of the R&D program. So it's not free money by any means. And the dollars are always matched 50 percent by the company."

> While government support for R&D is very important in the start-up phase of a firm, it is also welcomed to support long-term development.

Mackie says he doesn't speak for Newbridge on the subject of government funding — but personally, he's all for it as long as it produces results in the long term that will benefit the Canadian economy and boost the high-tech industry. "I don't feel apologetic about government-sponsored research," he says. "I think it's necessary if we're to keep people in Canada with things to do and jobs — we might have great universities, but if our children don't stay at home, we don't have a country either, do we? This is all part of the big plan to make sure that the brain drain doesn't go south, that the telecommunications specialists stay here."

Newbridge employed some 1,800 people in 1994, about half of whom worked in Ottawa. "That's changing, though," says Mackie. "We're hiring all the time."

The company has been growing at a phenomenal rate of 60 percent per year, and recently expanded into an old Mitel building across the street. A new building is under construction in the field adjacent to the existing Kanata headquarters.

Research and development spending have been fairly constant since the company's beginnings, averaging around 11 or 12 percent of revenues each year. "We split our R&D into two parts," says Mackie. "One part comes from funds we get from customers and from government grants, and the other part is ploughed back in from the profitability of the company."

> Business is shifting away from the mature markets of the US and Europe to the growth markets of South America and Asia.

Roughly 60 percent of the company's business is in the Americas, and the rest of the world divides up into 25 percent Europe and 15 percent Asia-Pacific. "Probably this year will be much more strongly Asia-Pacific, a 2 to 3 percent shift over the year," says Mackie. "But eventually we believe that Asia-Pacific will be a growth region for us that will rival what we've seen in Latin America in the last few years. About 25 percent of the sales in the Americas are actually in Latin America." The world market for multiplexers is worth $1.2 billion, and is estimated to be growing at a rate of 30 percent each year.

Product improvement and research and development are big priorities right now. Newbridge recently opened research facilities in its Virginia office to concentrate on local area networking technologies, and it's spinning off some of its more mature products into some supporting R&D activities in Europe, although the bulk of the research is still done in Ottawa.

Newbridge's biggest customers now are telephone companies around the world who are facing increased competition due to deregulation of their industry. To stay alive, these companies have been forced to improve their technological offerings. Where once their corporate customers were content with the benefits of conference calling, they're now demanding to be able to share everything else, too — spreadsheets, videos and other pertinent computer data — from thousands of miles away. Newbridge supplies telephone companies with the equipment they need to do this.

Newbridge knows it's pretty hot stuff these days, and it has a few good reasons for thinking so: it has up-to-the-minute technology; it has carved itself out a niche with huge growth potential and few serious competitors; it has struck deals with major phone companies for selling and distributing its products; and so far, management has done a sound job of keeping runaway growth reasonably controlled.

Setbacks and Mistakes.

Events in 1990 touched off a phase that left Matthews hostile towards the media for a long while. Articles began to appear with headlines like, "Newbridge: Choking on its own success?" and leads like, "Terry Matthews likes to yell," or "Has Terry Matthews stumbled again?" The articles documented in painful detail the bugs and glitches customers were experiencing with Newbridge products. Some of those customers — some of the biggest and most crucial — refused to pay their bills. Lawsuits followed. Short sellers circled the company like vultures waiting for their prey to weaken sufficiently, and profited hugely when the company's stock took a nosedive, dropping by about two-thirds. Newbridge's image suffered further blows from the resulting layoffs and cutbacks. Financial and trade publications sprinkled their articles about Newbridge with skeptical quotes from analysts, questioning both the company's staying power and management acumen.

It was clearly time for some aggressive damage repair. Matthews stuck tenaciously to his guns and got his message out: he still believed firmly that it was impossible for a company to make a go of it in the huge telecommunications market without taking a few tremendous risks. With the multiplexer market expected to be worth $5 billion by the mid-1990s, there was no room for second-guessing. Matthews is described by colleagues and employees as an optimist whose confidence knows no bounds. Mackie contends that Matthews's apparent aversion to the press stems only from an altruistic concern that articles always seem to focus too narrowly on him without paying adequate homage to the hundreds of employees who helped build the company. He concedes that this period was bad for Newbridge in many ways, but pins the blame quite squarely on the shoulders of the short sellers, whom he accuses of spreading sensational tabloid-style stories about what was really going on at Newbridge.

> It is often difficult to separate reality from perception when a firm with spectacular performance gets into trouble because of the money at stake. This is exacerbated in the high-tech sector, which is considered to be volatile in any case.

"There are a lot of causes associated with that point in time," he says. "The chief amongst them was we had expanded dramatically as a result of demands from our customers. A lot of costs were associated with the expansion that was going on. We had product introduction problems — not problems that were killing us, exactly, but as the product had expanded it became necessary for us to build it into something with a more cohesive hardware and software development. We had to spread the product family wider, we had to spread the company wider, the costs went up, and delays were incurred in getting the product out.

"We were listed on NASDAQ at the time," he continues. "It's my personal belief — though it would take some research to prove this — that around that same time, we took some major hits from short selling activities. As soon as there was any weakness detected — because we'd only been on the market for a year — the short sellers got after us and started spreading *National Enquirer*-type stories about what we were doing." That mania fed itself, Mackie says. "People couldn't believe any company could grow as quickly as we were doing. They thought we were snake oil salesmen or something. That's the way these people trade. The more they can persuade you or me or anybody else that this company may not be what it's made out to be, the more money they make. That's the whole concept of short selling."

That disaster struck just around the same time Newbridge was growing rapidly, massively expanding its product line. The short answer, says Mackie, is no single cause led Newbridge to the downturn it suffered in that period.

To recover from it, the company put Peter Sommerer — the former executive vice-president and chief operating officer, and an ex-Mitel employee — in charge of the company as president in 1991, and Matthews became chairman. "Peter basically tightened all the windows down — travel was restricted, expenses were restricted," explains Mackie. "I don't want to take anything away from Peter here, but there was nothing terribly magical about the business technique. He took charge and did it. Then we put a lot more focus on the planning and evolution of the product. Not that that wasn't being done before, but there were fewer processes in place, and what was happening was that — as is true with any entrepreneurial operation — in the early days you don't have a lot of process because you've got big things to be done. You have people who typically are not very process-oriented — they're more project-oriented, and they understand what needs to be done and they get on and do it, and the paperwork suffers. Every new company has to go through a transitional stage, from being heavily driven by people with a vision and an ability to work without any organization, to where you have hundreds of people and you just can't function that way any more.

Troubled times usually bring management changes.

"You can't have a company of 2,000 people operating willy-nilly," says Mackie. "There has to be some structure to all of it. The imposition of that structure was Peter's job."

Risks and Choices

Identifying Newbridge's own peculiar risks, says Mackie, is a problem. "Of course there are risks. There are all the risks of regular business, I suppose," he says. "But specifically, I see us being in a business that's maybe not risk-proof, but certainly risk-resistant. I think in the early 1990s and probably for another decade — and nobody can see beyond that — telecommunications is the technology that will be growing the same way computers grew in the '80s. The computer people had their day in the '80s, and the communications people are having their day in the '90s."

The risk of a technology shift that would make the kinds of products Newbridge designs obsolete, says Mackie, is almost nil. "We pretty much have all technological directions covered," he says. "I don't think there will be any technological changes or surprises.

There's no one out there with a better way of doing things. There's no major shift in the industry looming. It's possible that we could suddenly decide not to move as fast as we have, but that's a wild, outlandish idea — it won't happen."

Also, says Mackie, as far as competition goes, acquiring Newbridge's knowledge or technology is almost impossible. "It's not available in textbooks. It's not common knowledge," he says. "No one's going to graduate from university and decide to start manufacturing products identical to ours. It would be like deciding to just take up nuclear physics. We're in a specialized area where we have only a small handful of competitors to worry about.

"We're pretty unique in Canada," he concludes. "We're in a good, safe niche. I can't imagine, and I don't see, any major threats to Newbridge."

> In the high-tech game know-how is key. Technology is changing so rapidly that patents are only of limited value. That is why skilled people are so important.

The Outlook for the Future

In 1992, Newbridge invested in ATM technology in anticipation of a new telecommunications wave it expects to ride to success in the future: video-conferencing. The company is also poised to leap ahead with new technology if wireless satellites become widely used. It sees itself possessing a captured market, where the more you have, the more you need.

Increasingly, says Mackie, Newbridge's customers will be large telephone companies trying to offer better service to subscribers. But he doesn't see the company losing its initial base of private network operators, either.

"There's an interesting push and pull," he says. "You could imagine that trend for the future because telephone companies have a right to sell this communications capacity. That is their business, and they need the tools to do that, and hopefully they'll keep coming back to us to get those tools."

What interests Mackie most is that Newbridge is in a business where the cost of communications services are decreasing at a precipitous rate. The same sort of technological pushing and shoving that went on in the computer world, where computers became more and more powerful at less and less cost, says Mackie, will be applied to communications capacity in the 1990s. "Less visibly, but with much consequence," he says. "So a communications channel that costs $1,000 today, three or four years from now will deliver 10 times the capacity for $200. It's the same forces at work exactly. Since that's what's happening, the question becomes, where does the telephone company fit into all this? Those companies are usually very large, and usually that means they move fairly slowly. End users of telephone companies get frustrated with the slow speed, so they go out and seek ways to make better use of what is available. And I believe they will keep coming to us for that kind of equipment."

The fact that Newbridge sells to both sides of the street — to both private and public networking operators — means that as it continues to sell to private network operators, it will pressure public providers for the same kind of service. "It would therefore be in our best interest to continue to sell to both, and not ignore the private networking side," says Mackie. "The private networking people provide all the stimulus for us to do the leading-edge research we do on our products. Telephone companies are by definition more conservative. They want proven interfaces, proven techniques."

The risks are therefore quite different. As a consequence, the 70–30 ratio should hold for some time — 70 percent being telephone companies and 30 percent private network operators. "It's a difficult thing to nail down," says Mackie. "It depends so much on technology, on business opportunities, on the speed at which telephone companies can adapt to selling what they're selling now versus what they were selling years ago. There are so many variables."

The Competition

An advantage Newbridge has over its competitors, says Mackie, is the broad spectrum of solutions it offers its customers. "We have a wide range of competitors. Our competitors in one area are not seen in other areas. Whereas Ford can say their competitor is GM, Newbridge is a company that doesn't just make automobiles, but also makes buses, tanks, trucks, and bicycles. We don't have a competitor who does everything we do."

The most important strategy for beating out the competition, says Mackie, is managing the networks. "The fact that we can provide management through a wide range of technologies is a very compelling selling point. For example, it's well known that people who install local area networks for, let's say a million dollars, might spend a quarter or a third of that money providing the management support for that network. That's an awful lot of money — it's people, equipment, test equipment. Wherever your office is, you'll need some of these people. So obviously someone who comes along and says, 'You don't need all those people and all that equipment, we can do it centrally,' will get a lot of attention.

> Providing management services with the product cements the relationship with the client.

"That's the kind of thing we provide," he says. "It means companies don't have to hire all these experts with all their equipment, who if everything is working as it should, won't have much to do anyway."

Competition in Newbridge's market is fierce. In the old days, Newbridge would have had to contend with maybe three or four competitors making T1 multiplexers; now, there are more like 25 making T3s. Matthews has said Newbridge competes by maintaining a two-year research and development lead over its competitors. (The chief competitors are Timeplex Inc., owned by Unisys of Detroit, and Network Equipment Technology Inc., whose products are now marketed by IBM.)

Although Newbridge's biggest customers are telephone companies, it also competes with these companies by selling its technology directly to their customers. Telephone companies, for their part, are fighting each other and Newbridge to win back the private-business network market — Newbridge's second best customers.

"Our customers are either private network operators, or they're telephone companies," says Mackie. "We sell to both. That's the unusual thing about us as a company — we have competitors who sell to the private networking side and competitors who sell to the public networking side. The characteristics of our products as they relate to the standards that the telephone companies use to build networks are ideally suited to both groups of people. So we have about 100 telephone company customers around the world. But we also have tens of thousands of other customers — ranging from people who buy a couple of our products to customers like Bank of America, that sort of thing. All of the big Fortune 500 companies are candidates."

Also, he says, very small operations people who run "very skinny digital networks" that would link, for instance, a lawyer's office in Chicago to a lawyer's office in New York, are good candidates. These are people who would actually lease the channel between these two locations, and on that channel, using Newbridge equipment, they could send any combination of voice/digital signals they wished. "So it's not just for phones — it's for the widest range of services, including high-speed fax machines, transfer of large computer files," says Mackie. "It varies depending on what the company is doing."

Management Strategy

Newbridge attributes much of its early success to a solid distribution system and deft handling of international markets. Plain old workaholism is also part of the formula. Enthusiastic employees have been quoted in the past about how many hours a day they work — and they weren't complaining; they were boasting about their eight-to-six jobs that were often followed by seven-to-midnight evenings and sometimes weekends. Such dedication is fostered at least two ways: most employees own a good share of the company themselves, and would love to see healthy returns on their investments; and then there's the sense of family that Matthews reputedly works at generating. For instance, in 1988, more than 400 employees chartered a train to Toronto and spent the weekend there. There are regular information sessions, and Matthews, in less busy years, made a sincere effort to be available to anyone. In 1989, Matthews decided to build a fully staffed guest house for out-of-town employees, on site. Then there's the legendary trip he sent all of his employees on after the first year-end results: he flew 300 employees and their families to Florida.

> Taking good care of people, the major assets of a technology firm, is key.

"What's more important is the story behind that story," says Mackie. "An awful lot of people worked very hard with great sacrifice to their families before that happened. The families, the husbands and wives of the people who were working here, put up with a lot of trouble. It wasn't so much a reward for the employees as it was for the family. There were lots and lots of late hours and difficult times. Lots of people — my wife included — came in to help out at the end of the quarter with things that had to be done. This happened for several years — at the end of every quarter, people were coming in late to move cardboard boxes or put things in boxes or learn how to run the automatic test equipment so they could just keep loading the machines, loading equipment up. Children, wives, husbands — everybody came in."

One of Matthews's strategies early on was to concentrate on smaller networks, which were an easier sell because they were cheaper, and represented less of a risk for the first-time buyer. These networks were just big enough to link branch offices or several dozen computers. Matthews's reasoning was that once customers had these small systems in place, they would be so happy with Newbridge that when they needed something bigger, Newbridge would get their business. Four years after it was founded, Newbridge had sold more than four times as many multiplexers as its closest competitor.

Mackie finds the transition from entrepreneurial to "professional" management interesting. "Professional managers almost by definition are process people," he says. "That often means they like to manage the opposite of the kinds of projects that typically drive entrepreneurial companies. That transition is a delicate move — from everyone knowing what they're doing and which way they're going while they're doing things, and being caught up in that culture and that excitement, to an environment with less excitement and more process.

> The project orientation of the entrepreneur is in sharp contrast to the process orientation of the professional manager. Making the transition can at times be difficult.

"You'll notice at Newbridge that we've retained a lot of the small-company style, with very little, or no ranking or power structures in the company," he says. "People treat senior members of the company pretty much the same way they treat anyone else on the team. You don't have to go through somebody's boss or assistant to get somewhere. Some things that big companies find very strange are still part of Newbridge.

"We only started in 1986," he points out. "That kind of entrepreneurial culture doesn't get lost in such a short time."

12

Dynapro Systems Inc.: Steady Growth through Alliances and Acquisitions

Dynapro is a $50 million company that grew more than tenfold in the last 10 years by forging a crucial strategic alliance with a major US firm, Allen-Bradley, and acquiring two businesses along the way.

Dynapro Systems Incorporated is a company whose history reads like one long checklist of projected successes neatly accomplished.

When he first founded Dynapro in 1976, Karl Brakhaus was 27, with a PhD in engineering physics from the University of British Columbia — a research whiz, but a management novice, with no business background to speak of. Now he's president and CEO of a company whose 1993 sales topped $40 million and whose staff has grown from 3 to 350. Even more impressive is the fact that Dynapro's progress — from a three-person consulting outfit to a healthy, mid-size corporation — has been a relatively smooth ride on the treacherous treadmill of innovation, with no serious disasters and few major setbacks.

Now based on Annacis Island in the Vancouver suburb of New Westminster, BC, Dynapro is privately owned. The company, which manufactures hardware and software for user interfaces and control in industrial and commercial automation and public access systems, got its start not long after Brakhaus got out of school.

A newly minted graduate of UBC in 1974, Brakhaus worked as a consulting engineer at Columbia Engineering International, a full-service consulting firm in Vancouver. When he was hired, he was one of 100 employees. When he left two years later, less than 20 were left, and the company was on its way out. "I was hired supposedly to head up the computer department they were putting in," says

Brakhaus. "What I didn't know was that they had reached the peak of one of several five-year cycles ... All the projects they were working on were either being finished or cancelled, so people were let go. It gave me a real taste of what that kind of market is like," he adds.

So Brakhaus, armed with new microprocessor technology, technical genius and some contacts in the field, found himself two partners — William Gunn and John Fairclough — and started up Dynapro as a consulting firm. They began by attracting a few consulting contracts. "We didn't have big overheads," says Brakhaus, modestly explaining why he never thought starting his own company was the huge risk others might perceive it as. "I didn't have kids, no one was married, there was no house, no mortgage — basically, we just didn't have a lot of overhead."

He has come a long way since those hand-to-mouth days, though he still operates in a world considered by many to be high-risk and high-stress. By 1989, *B.C. Business* magazine had voted Dynapro one of British Columbia's 10 companies of the future, praising it as a company that manages itself deftly in a world of high-tech products with short life spans — one of a series of "fast-track performers poised for major success in the next decade."

"First of all, there was an opportunity," says Brakhaus, describing the start-up of the innovation process at Dynapro. "The opportunity came about through technological change ... We saw this technology and applied it."

Dynapro owes much of its success to Brakhaus — to his willingness to take risks, combined with the judgment not to take foolish ones and a rare ability to anticipate both what the future holds and what he wants it to hold. Another ingredient in its success formula has been Brakhaus's conviction that companies need to pursue a global market; most of Dynapro's customer base is in the US and abroad, in Europe, Asia and Australia. It owes the rest of its success to the combined ingenuity of its founders, who built the business more on ideas and alliances than on investments.

> Technology firms have to think globally from the outset. While the Canadian market can act as a springboard, it is too small for most firms. The favoured market is the US, which is 10 times that of Canada and has the same business culture.

History and Milestones

For the first few years, company founders focused mainly on consulting work. Then — in the late 1970s, in the days when fully dedicated word processors were the latest thing — Dynapro started a project to develop a word-processing system. "It seems so archaic now," says Brakhaus, "but back then, that's all the machines did. The whole system was built around a microprocessor."

Eventually the three decided it would be interesting to create a display system that could be used for industrial process control — food processing, beer making, packaging and bottling lines, and film processing, for example. Perceiving a need on the part of companies who wanted an improved way for people and machines to interact, Brakhaus, Gunn and Fairclough decided to design one. The result was the 1979 release of the Dynapro GRAFIX terminal. Replacing hard-wired, push-button control panels with graphics displays based on microprocessors, this technology was the first of its kind on the market.

Brakhaus didn't see this new product line as much of a risk; Dynapro wasn't the only company in this market. "From an innovation point of view, it was no big deal," says Brakhaus. "There were other expensive systems available already." Then, in 1979, they came across a company called Intacolour that was selling graphics terminals. "We took one of those and modified it, created software and hardware to put into it, and made it suitable for use in an industrial environment, at a price that was suitable for that market," explains Brakhaus. By now, they had purchase orders for several systems with a number of terminals each, although these would be in the designing stage for about a year. They continued to do consulting work in general industrial control applications. By 1980, Dynapro was still a consulting company consisting solely of Brakhaus, Fairclough and Gunn.

In 1980, Gunn left and was replaced by another engineer, Bob Angus, and the company began to market the product. "We started demonstrating it at trade shows and so on," says Brakhaus. Their first customer was Molson Breweries in Vancouver, which set the scene for years to come: display products used in industrial control applications still make up the bulk of the company's sales.

In its current incarnation, Dynapro consists of three different corporate subdivisions specializing in similar but different areas. The company makes the controls that allow people to give computer-guided machines instructions. Specialties include real-time acquisition and graphical presentation of information (such as public information kiosks, airline baggage handling, blood analysis machines), and products allowing easy user interface with computer-controlled machines and processes (touchscreens, digitizers); the company's main industrial products are a line of factory-floor operator terminals, touch control screens and supervisory control software. These products are in demand in situations ranging from manufacturing, sophisticated control rooms, public access terminals and process control, and at the component level, in pen-based computers and digitizers.

In late 1982, after the release of its GRAFIX terminal, Dynapro made a sales pitch to sell peripherals to the two largest producers of programmable controllers on the market — Allen-Bradley and Modicon — and prepared for the flood of orders they thought would pour in. Unfortunately for Dynapro, Modicon came up with its own display system; and Allen-Bradley, a leading US manufacturer of controllers and related factory automation products, and subsidiary of worldwide industrial giant The Rockwell Group, had also been gearing up to build its own. At first, it looked as though what had seemed the perfect market niche for the fledgling Dynapro was going to be overtaken by bigger, wealthier competitors. But Dynapro had one advantage: speed. It would have taken Allen-Bradley at least two years to develop its own similar product, whereas Dynapro already had one. So in 1982, the two companies put together a deal in which Allen-Bradley would "private-label" Dynapro's technology.

> Alliances can be mixed blessings. They can bring in investment, marketing capabilities and distribution networks, but the larger partner can also dominate the relationship.

The alliance was cemented when Allen-Bradley bought 25 percent of Dynapro in 1983, and later boosted its equity stake to 50 percent. The main drawback of this alliance has always been that Dynapro's name doesn't appear on Allen-Bradley products; the company's future research will continue to be determined mainly by Allen-Bradley's needs. However, this alliance has been the keystone in Dynapro's ability to make rapid inroads into the US market, providing Dynapro with a greatly increased customer base and distribution system.

"We started shipping product to them in the summer of 1983," says Brakhaus. "And we've been doing that for the last 10 years." In the year following the A-B deal, the number of employees at Dynapro doubled, from 30 to about 60, and sales started to climb. The end of fiscal 1984 saw the company hit $5 million in sales — almost double that of 1983.

That looks easy in retrospect, but Brakhaus and several of his staff jumped through major hoops to get there: in the beginning, there were several similar companies competing for the A-B deal, and although Brakhaus thought Dynapro's presentation and bid went extremely well, A-B wouldn't give them the contract: A-B refused to buy equipment, manufactured by Dynapro, that used parts from Intacolour, the firm that provided Dynapro with the necessary parts.

"Just before Christmas of 1982, we were on the short list for doing the deal with A-B," says Brakhaus. "The president, as a final element in the decision-making process, decided to visit the short-list finalists. They flew here in their company jet, and we had a meeting in our little, dinky boardroom, did what we thought was our very best presentation, slides, a nice, darkened room ... We thought we did a pretty good job. We did all this, and at the end of the presentation, they told us, 'Well, guys, we wish you all the best, but A-B will not have a product based on a product built by Intacolour.' So there was no deal."

This was because Intacolour, says Brakhaus, had for some reason built up a reputation for manufacturing somewhat unreliable products — for making display terminals that failed. "And we were using them as a base," he says. "We made them more rugged, actually, but it still didn't cut any ice with A-B.

"We were devastated," says Brakhaus. "We had a viable business even without A-B, but we really thought this would help us to dramatically improve our business." Then an engineer in product development came up with a suggestion that would help them get rid of any sign of Intacolour. "Over the Christmas holidays," says Brakhaus, "this guy took the chassis out of the Intacolour (a card cage with components in it) and put it into a big blue box and built a separate system — a card cage, which we put into a standard chassis, and mounted the disk drives in it. Over the holidays we actually prototyped a display generator that would drive a separate monitor.

"Then we phoned A-B up after New Year's and said, 'Hey, we have a product that doesn't use Intacolour.' It's probably fair to say they were quite astounded. We flew it down in January, and did the deal.

"After that we had to redesign everything — all the circuit boards, the chassis, everything, to get a product out. And we did it in six months — designed all our electronics, complete metal box, the software to run it — converted it all. By the middle of summer we were shipping products to A-B."

Selling to A-B became Dynapro's main preoccupation for about a decade, during which sales grew slowly but steadily. Recently, however, the company has begun to diversify, branching out in different directions by acquiring other companies. Within a year and a half — between 1992 and 1993 — Dynapro bought divisions of two other companies, expanding into non-industrial markets. Sales climbed gradually to $10 million in 1989; by the end of 1990, they had skyrocketed to $20 million. Dynapro acquired divisions of two new companies in both 1992 and 1993. The company seems to have done a sound job of absorbing these acquisitions: 1992 sales of $27 million climbed to more than $40 million in 1993, and Brakhaus is projecting sales of $50 million for 1994. "And we're going to hit it," he says.

> Acquisitions can often be an appropriate strategy for growth.

Dynapro's growth was incremental — a one-day-at-a-time existence — until about 1983, says Brakhaus, when the breakthrough finally came in the form of the alliance with Allen-Bradley. Dynapro was already building peripherals for Allen-Bradley products, and the larger company offered marketing channels to the US and beyond. For years after that, Dynapro designed products only for Allen-Bradley and diversified its products to meet the needs of other companies within the Allen-Bradley group.

By 1985, Gunn, Fairclough and Angus had been bought out of the company, leaving Brakhaus to continue to head it up with the Allen-Bradley team. Gunn had gone on to develop generic floppy disks that would replace the expensive pre-formatted ones then being sold by various computer manufacturers; Fairclough took a vacation from the industry for a while; and Angus started up another high-tech company of his own called Trionics.

In the summer of 1990, Dynapro was one of three Vancouver companies awarded substantial sums of money as part of the Western Economic Diversification Fund (WDF). At about $2.7 million, Dynapro received the most of the three companies. Around the same time, it received an additional loan of almost $2.5 million, supplied jointly by the BC Ministry of Development, Trade and Tourism and another (federal) WDF loan.

The company applied the money towards the $10.7 million cost of moving from its leased premises — split between South Vancouver and Richmond — to its present location in the new, 92,000-square-foot facility on Annacis Island in New Westminster. The company celebrated the official opening of the $7.25 million Annacis Island plant in September of 1991. Within three months, managers had noted an improvement in shortening the time the product took to move from the development phase to manufacturing. Dynapro products usually have a two- to three-year life span, and require one or two years to develop and begin production.

That money also helped develop eight computer-based graphic hardware and software products for the industrial automation market. The company credits these government contributions for its firm place in the Canadian high-tech scene — this money is part of what enabled it to remain in Canada instead of joining the mass exodus to the US in search of cheaper operating costs and lower taxes.

Meanwhile, in 1991, Brakhaus himself was awarded the Cecil Green award for entrepreneurial science, bestowed by the Science Council of BC and named after Sir Cecil Green, who also attended UBC and later founded Texas Instruments, an early US leader in the development of microelectronic circuitry.

In 1992, Dynapro went after ownership of a touch control screen product line. The company was again awarded an interest-free $3 million loan from the WDF, which it put towards the $10 million needed to acquire the technology for touchscreen computers. This marked a change in strategy for the company, which until then had focused mainly on developing new products. The new line, executives reasoned, would add more commercial business to its existing customer base. Only months after this decision, the company purchased this product line from the John Fluke Manufacturing Company Ltd., based in Everett, Washington, for $10 million, one-third of which came from a WDF interest-free loan. ("We went cross-border shopping and we brought home a manufacturing line," Brakhaus said at the time.) The remaining $7 million came from Dynapro's internal cash flow and a loan from the National Bank. Within four months, it had integrated the touchscreen terminal business, extended its product line and expanded sales activity to meet new customers.

> Governments use a variety of incentive programs, not only to give firms a competitive edge in developing products, but to offset comparative advantages offered by other jurisdictions.

"We started shipping display products to non-industrial customers," says Brakhaus, "with our own logo on it rather than Allen-Bradley's." Sales and revenue growth led to the hiring of more than 50 new employees in 1992, bringing the total staff to 230.

Brakhaus accepts this success as proof that Canada can compete in some areas of manufacturing without relocating to the US. The acquisition, and the continued success of Dynapro, caused Michael Wilson, then Minister of Industry, Science and Technology, to say in 1992 that Dynapro exemplifies "the kind of company Canada needs to compete internationally." Brakhaus attributes the company's success to a consistently increasing demand for Dynapro's products, which enhance the user or human interface in automated systems. "Because Dynapro provides hardware and software that improve production efficiencies, our products are in demand regardless of overall economic conditions," said Brakhaus at the end of a hugely successful 1992.

In 1994, about 75 percent of Dynapro's sales were to Allen-Bradley. The remaining 25 percent, which included a wide array of non-industrial customers, was due largely to recent acquisitions, which in turn had allowed the company to hire another 100 people in 1993. That year, Dynapro also purchased the Thin Film Products (TFP) line of W.H. Brady Co. in Milwaukee for an undisclosed price, the company's second diversification effort involving a US business. The purchase included the assets, technology, manufacturing plant and customer base of the TFP unit. Dynapro supplies touchscreen terminals and software, while the TFP unit makes such products as touchscreens for computer-driven displays, controller electronics and pen-based digitizers. The product line is sold mainly to manufacturers of pen-based computers and touchscreens in industrial, medical, telecommunications and retail applications. TFP unit customers include GE Medical, Oregon Scientific, NCR, Nautilus, IBM and Hughes Avicom. Publicly held, Brady had 1992 sales of US$236 million.

Setbacks and Mistakes

Although in retrospect Dynapro's record appears unblemished, the company has not escaped without a few minor setbacks that seemed like nightmares at the time. "The grass always looks greener in retrospect," says Brakhaus. "There have been endless disasters. There are always crucial decisions to be made, where if you make the wrong one, you can hurt yourself very badly or go out of business."

This can happen especially in periods of rapid growth or transition, when sufficient cash resources are essential in sustaining and managing that growth. One very close call in 1990 taught Dynapro that lesson.

"Before the IBM PC came out, our product was one where we designed the computer itself, the entire system," says Brakhaus. "When the IBM PC came out, competitors started making products built around that computer. Demand for that kind of product went up, and we didn't have one. So we converted our product so it would run on a PC.

"Our sales of that new product were zero for six months. In hindsight, we could see why that happened — for one thing, the sales force basically decided that the original part we'd been making was obsolete and that therefore there was no point in selling it. On the other hand, the new product, it turned out, wasn't rugged enough for use yet. We had accepted it as being adequate, but as it turned out, it wasn't. So the existing product, which had enjoyed reasonably good sales, was suddenly obsolete; people stopped buying the old product, but they didn't start buying the new one, because of unfamiliarity, the lack of hardware to run it on, things like that," says Brakhaus.

"Luckily we had another product we were still shipping, and those sales, for some reason, went up. Still, we nearly drained our cash resources in six months. We recovered from it, but it was traumatic — a really close call. If we hadn't had cash resources accumulated, we would have been dead — another statistic."

The company was forced to lay off 18 workers that year. Its sales figures were still solid by the end of the fiscal year, but only after a shaky recovery that nearly wiped them off the map.

"Deep pockets" are needed, not only to develop new products, but to cope with the volatility of the market place.

Risks and Choices

One issue Brakhaus has wrestled with involved choosing the most economical manufacturing site for Dynapro's products. He's a firm believer in supporting Canada and the local high-tech scene by manufacturing locally. Competitors in his field, however, believe just as strongly in the value of manufacturing offshore — Hong Kong, for instance, where consumer-oriented products with small electronic components and major labour components can be made much more cheaply, sometimes saving the manufacturer as much as 30 or 40 percent of the costs. Dynapro's products, conversely, tend to be low-volume products that are changed on a regular basis, making foreign assembly not only undesirable from an ideological viewpoint, but also difficult from a logistical, economical viewpoint. "If someone builds it for

Manufacturing in countries with lower labour costs is not always the answer. Some of the hidden costs associated with long distance communications, differences in languages and business cultures, for example, can still make Canada the preferred site for manufacturing.

you, they acquire the technology," Brakhaus points out. Then there's the cost of communicating with the manufacturer, and language and cultural differences — complex situations that require careful consideration. So far, Dynapro has avoided them.

A second issue is the fact that most of Dynapro's revenue is generated by sales to Allen-Bradley, which means they have less-than-ideal control over both revenues and product migration. As Brakhaus puts it, "Our fortunes depend on how Allen-Bradley's fortunes go."

The Outlook for the Future

When Brakhaus looks to the future, he sees Microsoft Windows. Now that manufacturing and administration are cohabitating on Annacis Island — Dynapro's own products monitor and control the temperature, lighting and security — the company is set to take off on another growth spurt. "Microsoft Windows is a whole new transition we're involved in now," says Brakhaus. Having learned from the near miss of 1990, Brakhaus says Dynapro, although excited about this newest transition, will not rely upon it exclusively or sink major funds into it.

For Brakhaus personally, management now dominates his schedule. But by no means would he classify himself as a manager at the expense of remaining an entrepreneur. Risk-taking, he maintains, is still top priority in his job description, as long as it's not taken to foolish extremes.

"We build on strength," says Brakhaus. "We haven't taken wild jumps into the blue. Even our founding product was not really, truly innovative — it was the application of an existing concept in a different niche, making money out of it. We've been innovative in the small things, in how we do things." In 1993 Dynapro was projecting 1994 sales of $50 million. The acquisition of TFP has seen its sales double within the year of purchase, and Brakhaus expected sales to double again in 1994. About 15 percent of all sales — sometimes more — will continue to go back into R&D, a strategy Brakhaus maintains has contributed to the company's impressive growth curve. About a quarter of Dynapro's total staff — 90 out of 350 — are dedicated to product development.

> Technology firms have to spend a lot of money on R&D, usually more than 10 percent of sales, if they are to remain at the leading edge and support their core technologies.

The Competition

Historically, Dynapro's main competitors have been Allen-Bradley's competitors. Now that the company has branched out, they have to watch their own market niche more carefully. "In the terminal market, for instance, we're competing with companies like Zicon, Neutron, TCP. But these are not really large companies. They're no bigger than us, certainly," says Brakhaus. "In the software market, we still sell through A-B; there are other companies that sell directly into that niche as well. But they're our size or smaller."

Market share depends on the product — Dynapro has all of A-B's business, "if you look at it that way," Brakhaus says. That percentage actually translates into a 40 to 50 percent market share of A-B's customer base. Of course, as Brakhaus points out, "The more broadly you define the market, the smaller your market share gets.

"We're not selling into niches that are really, really clear," says Brakhaus. "These products have wide-ranging uses. You can't even define them by the type of industry."

Management Strategy

Early in 1990, Dynapro awarded $10,000 worth of money prizes for outstanding achievement in areas like innovation in process design, quality, hardware or software design and excellence in product definition. The goal: to reward employees for keeping the company competitive. In a three-month period that ended in early 1990, the company's product rejection rate dropped to 2 percent from 14 percent, after which the entire manufacturing staff were awarded company jackets featuring the Dynapro logo and printed with the words "world-class manufacturing team." The 10 cash awards presented in January of that year ranged in value from $850 to $1,500. "We chose the awards to reward the kinds of things we wanted to emphasize," said Brakhaus at the time. "We are competing in a global market place, not with just a local company up the street."

Brakhaus's management style emphasizes communication, incentive and quality. Brakhaus himself attributes it to an instinct for self-preservation, pointing out that as an entrepreneur, you can deal with situations on your own, but once the company grows, you have to have people around who are capable, motivated and happy to help you deal with these situations. He also attributes Dynapro's success to knowing who to hire — that, he says, is how they made the leap from academia to business without nose-diving. Equally important is the ability to assess the market place and know what's needed.

"Rapid growth is a very stressful time — that's when a lot of change occurs," says Brakhaus. "If you're not capable of coping with change, or envisioning change, or anticipating what should be at the end of that change, you may make a lot of wrong decisions."

Previous management experience is no prerequisite for success, says Brakhaus, who started with none himself and built up his business with two partners who also had none. "If you're used to running a very stable company with set procedures, that experience may not help you. It might even hurt you — you might have preconceived notions about how things should be, and they might be totally wrong.

"What we've certainly seen in the last decade is a lot of change. And Dynapro has gone through all sorts of stages of growth, from absolutely nothing up to a small-to-mid-size company. We've seen changes in administrative systems, procedures, approaches to developing products. The technology we use now is completely different from what we used 10 years ago.

"The senior management team right now had very little experience when they joined. They've grown up in the company, with one or two exceptions."

The company won't see a shift from balancing on the razor's edge to sitting on its managerial laurels. "We're still inventing all the time," says Brakhaus. "Of course, it's a different kind of entrepreneurship than when you're three guys subsidizing yourselves. What we do now is different, but it's still risk-taking. We're always trying to push the envelope, but not be silly about it. You want to take calculated risks, risks where you have a fair chance of succeeding — pushing the limits, but not too far, knowing that if you really bite off too much, the outcome can be quite drastic. You have to be fiscally responsible and fund yourself."

He plans to continue to be aggressive about conquering new markets. "We haven't sloughed off at all," he says. "We also have 90 people in product development, and the range of products is growing."

An attitude of persistence, a talented pool of people, morale-boosting steps like rewarding achievement, and reinvesting 15 percent of sales back into research have all contributed toward the company's growth.

"You have to be able to anticipate what's going to happen next, be flexible, open to change. This industry is driven by change," says

Brakhaus. For Dynapro, that has meant setting up a complete sales channel of its own — something it never had before.

"Last year, for instance, we engaged in some major change. For 10 years we were building and manufacturing products for A-B only. Now we've acquired two other businesses, and that changes our business considerably. We're now dealing with sales organizations in Europe, the US, South America, Australia, the Far East — all outside of A-B."

Brakhaus also cautions against the too-rapid growth that can sometimes accompany truly revolutionary technology. "You have the possibility of really going from nothing to something big in a hurry," he says. "But we've opted for pretty steady growth. It's steady, but it adds up."

Dynapro's greatest strength, says Brakhaus, is a formula that includes culture, people and a solid technology base. "They all come together," he says. "The people are good, and they're working on a solid base. We have good commercial relationships set up; we have the system set up to take us forward. We have a culture that I think is appropriate. And we've got good talent. You have to have a good balance between all these aspects, and you have to have them all focused."

Setting up a sales channel is the big new challenge. "We could still have a major disaster, although I think we've covered all our bases really well," says Brakhaus.

The real test of any company is how it deals with a looming disaster, says Brakhaus — for instance, choosing to lay off 18 people, as Dynapro did once, instead of sinking into an unrecoverable situation and going bankrupt.

How you deal with failure is also a good measure of entrepreneurial success, says Brakhaus. "Don't expect to succeed the first time. A lot of little things have gone wrong at Dynapro along the way, but we got past them and continued to grow. Some of the most wealthy people in the world actually have a bunch of failures in their background, but they kept persevering. If you're entrepreneurially inclined, you really have to learn to deal with failure.

"It also governs how you think about things, how you deal with things, how aggressive you are, because if you're really concerned about failure, you probably won't take the risk in the first place. You may hold back at just the wrong time, or not take the risk it was necessary to take, and consequently fail. Entrepreneurship is all about risk, and taking risks, taking gambles."

Also, be ready to do anything. "As this company has grown, I ran manufacturing, I designed products, wrote software code, I did the books, wrote my own letters, still do, sometimes. You have to do a bit of everything," says Brakhaus.

Above all, be keenly interested. "You'd drive yourself nuts," says Brakhaus, "if you spent the kind of hours that you have to spend to do this, if you really didn't like it."

13

Cognos Inc.: Balancing Entrepreneurship and Professional Management

Cognos is a $150 million software company in the process of renewing itself by changing its base technology. To do this the firm has had to restructure its management so as to be better able to jump to the next S curve.

Cognos Incorporated, now the largest independent software company in Canada, has often been documented by the popular press as having a management style more conservative than that of most similar high-tech ventures. But its chairman and founder, Michael Potter, disagrees adamantly with this characterization. According to Potter and his long-time colleague and current vice-president of technology, Bob Minns, Cognos has changed.

The success of Cognos today is largely due to a decision Potter made shortly after he got out of university some 20 years ago. Now chairman and chief executive officer, he bought out his two partners when Cognos (then called Quasar Systems Ltd.) was still a small consulting company.

The original partners, Alan Rushforth and Peter Glenister, first founded Quasar in 1969. Potter joined them in 1972, a recent graduate of the University of British Columbia with a master's degree in physics. His background was operations research — the science of solving business problems with applied mathematics. Within a year he had become a partner, and in 1975 he bought out the founders, becoming president. Along the way, in 1973, he convinced Bob Minns to join the company. The two had met prior to 1972, when both were working as consultants for the federal Department of Supply and Services.

The partners sold their share of the business to Potter because their opinion of how to run it was quite different from his — they preferred the small, 30-person outfit where everyone concerned could make a good living without too much stress. Around 1979, under Potter's management, the company started to veer away from consulting and into the packaged software business. In 1983, Potter changed the company's name to Cognos Incorporated — from the Latin *cognoscere,* to know. The company went on to become a world leader in applications development software, and is now the largest company of its kind in Canada, employing more than 1,000 people in 50 countries worldwide and taking in more than $150 million in sales annually. Its mandate is to develop, market and support advanced client/server development tools and reporting applications on a wide range of open and proprietary platforms, including UNIX and Microsoft Windows.

> Making a major transition from consulting to selling software products sets a new direction that requires leadership and perseverance.

Potter's decision to buy out his two founding partners was only the first in a series of executive decisions that resulted in management changes, each initiated by Potter, always in anticipation of upcoming periods of change, or following periods of serious financial loss. At one stage, Potter thought the growing company needed experienced, IBM-style management, so he imported an ex-IBM manager, who lasted less than three years. By the time he left, Cognos was suffering serious losses. Potter, who today is still careful to highlight only the benefits this manager brought to Cognos, sat back down in the president's chair to craft a plan for damage repair. Soon after, he brought in two new managers — one to be president, and another to oversee research and development. That president lasted two years, and was eventually replaced by the VP research and development, who still has the job. Potter's explanation is that each of these individuals had unique skills that were well-suited to Cognos's needs at a specific time.

Cognos's first software product was Quiz, developed by Minns as a tool for making Cognos run more efficiently. (Minns has a degree in geography, but says he always had an aptitude for math. He picked up his computer science skills on the job.) Reasoning that since the program worked so well for them — cutting work time down at Cognos significantly — the company began to market Quiz to other businesses. By 1984, sales of Quiz had earned the company close to $18 million.

The success of Quiz financed an expanding sales force and further research and development. Quiz itself spawned an entire family of tool kit programs called PowerHouse, which has become Cognos's flagship product. Most of the products developed since then have been based on PowerHouse, an advanced computer language designed to simplify programs for business managers while developing complex business applications for them. Its different packages simplify programming, so that what might take a month to do in Cobol could take less than a day with a PowerHouse package. PowerHouse applications are used in a wide range of industries, including banking, transportation, manufacturing, health care and defence.

Potter says Cognos really began to take off during the mid-80s. "We had some fairly good growth in the 1970s," he says, "but I think the sharp rise occurred once we focused the business on the product side, and the mid-80s were probably the sharpest growth rates. It was a strategy that opened a huge market to us, international and cross-industry. We had a very successful technology and a huge market that was worldwide."

History and Milestones

Although Cognos began as a consulting company typically for government departments, custom software design allowed the company to save the capital that would later finance its transition to a products-based company. They were interested in moving into other businesses as an original equipment manufacturer. "What that means is that we would buy hardware, say from Hewlett-Packard or Digital Equipment, and we would put together a software system and sell the system along with the hardware to the customer," says Minns. But they were beginners, with lessons to learn. "We would get a discount from HP and make some money on the software," says Minns. "But in reality it didn't really work that way. We weren't really very good at it. What would happen is we would say, 'Oh, we can buy this machine for $100,000, we can sell it to the customer for $150,000, say, so that's a $50,000 profit, so we can give the software away.' Well, it doesn't actually work like that."

Government procurement is an important instrument that helps the development of technology firms. Government can be a demanding buyer.

The first products actually evolved from the custom software development Minns had been doing for other businesses. "We had done several of these projects and we were clearly not as productive as we could be," says Minns. "And it was fairly clear that we could sort of systematize the development process that we went through to build these systems."

That finally happened when Minns was working for the National Sports and Recreation Centre in Ottawa on yet another Original Equipment Manufacturer (OEM) project. "We were losing money, as we usually did, and the way you manage these projects is when you're really losing money you just reduce the staff down to the bare minimum," he says. "The bare minimum is one person. I was now running this project. Because of the large variety of technical problems, I had to spend a lot of time in the evening at customers' sites just repairing the damage that had been done during the day. So one of those evenings I decided I would take some of the ideas that we had been developing, and systematize them. I actually developed Quiz — the first piece of PowerHouse — for use on the next project we were planning to do with this customer, in effect to make my life easier." He originally developed it without any thought of turning it into a product.

The next step, says Minns, was a presentation he made in Lyon, France, at a Hewlett-Packard users group conference. He remembers being surprised at the number of people who attended it, and at the level of response in the room. "Mike and I quickly came to the conclusion that we could turn Quiz into a product that we could sell," he says. "So I set myself the aggressive target of selling 10 copies in the first year for $100,000. We sold that in the first month. We did probably $1 million over the year."

That caused some problems for Cognos almost immediately, says Minns, because he quickly hired a team of people to help sell the software, and had no idea how to manage them. "I was a developer," he says. "I didn't know much about setting commission schemes and so on. These people ended up making more money than either Mike or I at the time."

The growing team at Cognos quickly realized that here was a business they could make something of, so they confidently took the next step, developing a program that would complement Quiz: Quick, which allowed users to design data entry systems.

"We followed the same process," says Minns. "We made another presentation of this product at another Hewlett-Packard users group meeting, got some more interest, and developed Quick further on a contract basis. The Northwest Territories government was a big user of HP products, and they really wanted the kind of technology we'd half-developed, so we worked with them to take it further, and implemented a generalized data entry system.

"We took that code back to Ottawa and turned it into a really good product called Quick. Later we put Quick and Quiz together — and it became PowerHouse."

A year after the company changed its name to Cognos — 11 years after Potter bought it — there were 300 employees. The company had become one of the largest independent suppliers of business software for minicomputers in the world. While its competitors were struggling to get to the other side of the recession, Cognos was sailing along famously, due largely to its focus on a new line of packaged software: off-the-shelf programs for businesses. By 1984, more than 80 percent of the company's revenues were derived from packaged software. The rest still came from consulting, a part of the business that was continually being scaled down to make room for designing and marketing new products. About 90 percent of Cognos's business came from outside of Canada.

"Clearly the early part [of the company's financing] was provided by consulting," says Potter, describing the progress of Cognos's financial resources. "We bootstrapped ourselves. We didn't require external financing. We had a cash flow from the consulting business, and we had the infrastructure in place — people, offices. This was still in the early '80s, around 1982. That was when we entered the second phase. We went to an established venture capital organization — Noranda Enterprises — and got significant financing, in the order of $2 or $3 million, which was a lot, considering the size of the company." Noranda is still a key shareholder as a result.

> As financing needs grow, firms have to go increasingly to professional sources of capital.

In retrospect, the company attributes its success during those years to conservative management, team spirit and a consistent lack of arrogance. These elements carried Cognos through the recession with flying colours, leading them to decide to go public in the spring of 1984. However, with most of an initial prospectus written up and all systems go, the move was stopped dead in its tracks by the stock market, which Potter and company decided was far too weak for high-tech stocks at the time. He decided the share issue price couldn't come close to reflecting the real value of Cognos.

No one was too upset by the postponed plans; Cognos was already making plenty of money, and had funding from the investment arm of Noranda, which owned 30 percent. Plans for going public were temporarily shelved, and didn't resurface again until 1986. Still, the postponement was a minor setback; the company had planned to use the money to build new offices, increase research and development, and expand marketing.

For most of the 1980s, Cognos focused on the mid-range computer market; most of its software products were geared towards systems like HP, Digital Equipment and Data General Corp. Towards the end of 1984, Cognos became involved in artificial intelligence research — the replication of human thought processes by computers. The company headed a $100,000 study, funded by the federal government, to try to determine what direction Canada should take in artificial intelligence research.

In 1985, Cognos, with almost 600 employees, moved into a new six-storey office complex in Ottawa and relocated its American head office from San Francisco to Boston, in order to tighten its link with Ottawa. Major inroads were being made into the Japanese market. But by 1986, expenses were starting to trouble the company's board of directors. The new office quarters — of which Cognos owned 50 percent in a joint venture with Campeau Corp. — were expensive, but most of their money was going into an expanded sales force in the United States, causing the company to sustain losses. The sales force in the US had doubled because of Cognos's efforts to market PowerHouse, its key product. Revenues rose by about 50 percent by the end of 1985, but expenses jumped 60 percent. Profits were evaporating, and Potter imposed some tough measures to control spending: a hiring freeze, along with cutbacks to capital expenditures.

In the middle of 1986, the company went through two important changes. It went public, raising about $12 million; and Potter decided to step down as president in order to bring in someone with more experience at managing a company the size Cognos was quickly becoming. He hired Thomas Csathy, who had been president of Burroughs Canada since 1982, following two decades with IBM. Csathy became president and chief operating officer. Both Potter and Csathy agreed at the time that the company needed more internal management strength if it wanted to play in the big leagues; Csathy's mission would be to keep Cognos's rapid growth under control. By 1987, Cognos had raised a further US$27 million with a public offering in the United States. It was now trading on both the TSE and NASDAQ, had 38 sales offices around the world, and was planning new ones in Scandinavia; but it called a halt to further expansion until a sure market could be secured. Cognos bought two small American computer technology companies that year, and began scouting for further complementary acquisitions in Europe.

Part of the reason for declining revenues in 1987 was Cognos's attachment to Hewlett-Packard minicomputers. In recent years, HP customers had been responsible for almost 60 percent of Cognos's sales; by the end of 1987, HP sales had fallen flat, taking Cognos down with them. Also that year, two American competitors — Oracle Corp. and Info Build — beat Cognos to the market with a fourth-generation language tool for microcomputers.

> Tying a firm's sales to one major client base can lead to vulnerability.

These events set 1988 up to be a less successful year. In the quarter ended November of that year, Cognos lost $1.8 million on sales of $18 million. Its stock had dropped 42 percent below the value at which it had originally traded just after going public. Sales were still growing, yet the company was losing money.

Csathy and Potter chalked these losses up to growing pains, rationalizing that the company was still working up to a management structure appropriate to a medium-sized company. One of the first changes Csathy had implemented upon his arrival was to combine the product development and marketing organization for each of Cognos's different product lines in an effort to streamline the business. He also added a direct sales organization to attract customers by phone and mail. In an effort to respond to changing market demand, he increased spending in research and development to 14 percent of revenues in 1987 and to 15 percent in 1988. Researchers began working on a new generation of software designed to write other programs for users with no programming knowledge.

Throughout 1988, Cognos continued to sell its products in about 50 countries. It had stopped losing money and was profiting from the introduction of several new products to the market within the previous year. Potter and Csathy defined the chief upcoming challenge as divining ways to attract new customers; tailoring software to clients' needs; helping non-professional computer users to gain access to data bases of other computers; and getting on-line access to customer information for sales reps. Cognos also started to gear up for a new line of products that Potter and Csathy suspected would be big winners in the 1990s: information systems, as opposed to computer languages.

Despite their lofty ideas, the company sustained unprecedented losses throughout the end of 1989 and early 1990 — a period that marked the beginning of the end for Csathy. Sales continued to climb, but profits kept right on dropping. In September 1989, facing a $4 million first-quarter loss, Potter was forced to begin some serious restructuring. Csathy resigned. Potter shuffled 20 employees into new positions and cut 20 more positions entirely. By December, following a further loss of $6.2 million in the second quarter, Potter found himself taking even more drastic measures. He eliminated about 100 jobs.

Poor performance usually leads to management changes.

It was becoming apparent that the skilled, schooled management techniques Csathy had brought to the company were no match for the enthusiasm of innovation and entrepreneurship that had put the company on the map in the first place. So, changing tactics, Potter brought in two new managers: Ron Zambonini, tempted away from American competitor Cullinet Software, became vice-president of R&D; and Jeffrey Papows, an R&D specialist, became president. (Zambonini was promoted to senior vice-president of R&D in 1990, and would become president and chief operating officer in 1993.)

In 1990 Cognos introduced InQuizitive, a terminal-based, menu-driven report writer that provides direct access to corporate data. It featured a friendly pop-up menu and report painter, so users could produce sophisticated, custom reports in a matter of minutes. That program was followed in 1991 by Executive Information Systems and Database Query Programs, designed to exploit the industry movement from centralized computers to individual computers linked on one network.

Cutbacks, management shifts and renewed determination helped Cognos get back into fighting shape again quickly. The list of accomplishments for 1991 included partnerships with a series of established companies internationally, from across Europe to Southeast Asia. The company branched off in a new direction, combining desktop and mainframe computers, which Potter predicted would be a dominant technological feature of the 1990s. Development of PowerHouse continued, growing more and more sophisticated. Zambonini is reported to have told Cognos's programmers that, as it stood, the PowerHouse program was "ugly." He compared it to a BMW — so beautifully engineered that its proud developers forgot it also needed paint to be aesthetically appealing. Researchers began sprucing up some of the key programs.

In 1992, Cognos started coming out with programs friendly to computer non-experts. One of the first such programs was Impromptu, designed to make it easy for "regular employees" to summon the company mainframe from their own terminals and request specific files, without having to phrase the demand in complex computer terminology. This marked a change in Cognos's target market, since for most of its history it had been selling to senior information systems managers. It didn't abandon its traditional market; it continued to sell its more complex software to its typical customers. But the new simplified software represented another rapidly growing market, accounting for 7 percent of sales by 1992 and expected to climb to about 50 percent in no time at all. Cognos saw itself entering another transition period, heading towards desktop tools and open-systems products. The plan was to develop a broader customer base among users who needed uncomplicated, quick access to corporate data. By now, PowerHouse was being used by about 20,000 customers in some 68 countries.

By 1993, Potter, still chairman and CEO, had identified two dominant trends that were influencing Cognos's direction. One was the demand by users to have access to corporate data — to be part of the information system in their respective companies, as opposed to being isolated on a single PC. The second was a shift towards open systems — in other words, towards applications that would outlive their environments, allowing organizations to exploit new technology with minimal expense. Cognos had developed a series of casual but crucial alliances over the years with companies like Hewlett-Packard, Digital Equipment and IBM; now it started to deal more and more with vendors such as Unix and Sun Microsystems. But the company has always remained independent, in the sense that none of its allies have ever owned a piece of it.

Cognos's newest products were expected to comprise as much as a quarter of 1993's sales, which perked up again after another management change earlier in the year. Zambonini replaced Papows as president, and Papows left the company. Potter says the change was an amicable one from both sides, necessary because so many different functions — including sales, corporate marketing, product marketing, finance and R&D — had to be more tightly integrated. He decided Zambonini had the most diverse, flexible set of management skills and would be the best candidate to iron out all the kinks. The plan for now is to stay as close to a grassroots entrepreneurial style as it is possible for a company with more than 1,000 employees to do.

Growing up, Cognos received no government funding. It finally got $8 million from the Ontario Technology Fund in 1991, which it matched, using the total to fund a project called Titan. The purpose of the project was to develop a new desktop division in collaboration with Carleton University, the University of Ottawa and the University of Guelph. "A lot of the desktop set, for instance, came out of that initiative, along with some of the improvements to PowerHouse," says Minns.

Potter credits the government with being a factor in Cognos's development in two ways: the government was a steady client for consulting in the early years, and provided research and development support in later years. "We use government money exactly the way I think it should be used: to reach further into high-risk R&D than we'd be able to go ourselves," he says. "You've got to have cost-benefit analysis even in something as difficult to quantify as R&D. So you're going to be limited in how much risk you're willing or able to take; you expect a certain return. When someone comes along and says, 'I'll match dollar for dollar' and you've got a very interesting project, then you start thinking about going beyond it."

> Government financial support is important, and at times critical, to technology companies, a fact that neo-classical economists often dismiss because they do not understand the innovation process.

The funding they've received so far, says Potter, was "not insignificant. I wouldn't call it funding that was basic to the development of the company, but it was funding that allowed us to reach into some high-risk R&D."

Potter characterizes Cognos as a company clearly in the mature phase of the major markets it historically has dominated. "The question now is, what's the next step towards growth?" he says. "This is what you're seeing in Cognos with the desktop and client/server market. It's a huge and very aggressive commitment to move to the next growth curve. That transition — which is a discontinuous shift to a new technology and a new market — is the story of Cognos today. What we're doing is aimed at shifting into that business. We're actually two companies on two separate curves, where one is very aggressively shifting its resources to the other."

> To jump to the next S curve, a firm first has to recognize that it has leveled off on the previous one. Many firms do not.

About 45 percent of Cognos's market is in the United States. Canada and the Pacific Rim together account for a further 15 percent, and the rest of the business is generated in Europe.

"People characterize our customers as the Fortune 2000," says Minns. "What I notice is that we deal typically, in our PowerHouse business, with the departments of really large organizations, or with medium-sized organizations. In other words, Cognos is probably large enough to be a customer of Cognos." In the Hewlett-Packard market, Cognos has cornered at least 60 percent of the market share, says Potter; in the desktop area, it already has about 50 percent; in other markets, Cognos's share is much smaller. "The Unix market, for example, is in the single digits," says Potter.

"In desktop tools, a strange thing happens," he says. "People think, 'Oh, this is a PC product, this is for individual users to buy.' But for the most part, the most important customers in that area are multinational organizations — people like Chrysler, et cetera. As it turns out, the most important customer for the small PC product is the big multinational."

Cognos's goal, says Minns, is to be one of the top three players in each of its niches. In some cases, that's already true. "Ideally we'd like to be the top player," he says. "We tend to define niches. We say, okay, we can be the number one supplier of application development tools. We are the dominant player in the niches we choose to specialize in." The most rapidly growing arm of the business is its new desktop division, which Minns says is entirely oriented towards selling packages that run on desktop computers. That division is growing at rate of 100 percent per year, but it still represents only a small percentage of the company's total revenues. "If you look at us and compare us with a lot of the new start-up companies [specializing only in desktop packages] that have just gone public, we're quite a bit bigger than them," says Minns. "And I'd say we're more profitable than them. But we're part of this much larger organization."

Cognos spends an average of 14 percent of its revenues on research and development, all of which is done in Ottawa. Currently, Cognos plans to market an executive information system and database query program. The idea is to keep up with the trend towards network-linked personal computers as they replace large centralized ones. Also in the works is programming designed to be used with Windows — a system intended to allow even the most computer illiterate managers to keep close tabs on their businesses. This is likely to result in declining revenues for the short term, while Cognos expands its existing product line to make room for smaller, desktop computer applications.

Demand is expected to increase for open systems, allowing the user to choose the most suitable computer for the problem being solved, for instance, or to choose whatever graphical user best fits the interface — and Cognos will have to move away from centralized, proprietary markets such as Hewlett-Packard and Digital Equipment. The biggest advantage of open systems is the freedom to change and expand without altering or compromising quality or performance. So far, Cognos has responded to this changing market demand with PowerWindows, PowerPlay, Impromptu and InQuizitive.

Potter says the transition Cognos faces now is similar in some ways to the transition Cognos made from a consulting firm to a products-based one in the early days — but there's a key difference: this time, the shift will be smoother, because instead of changing the nature of the business, he's gradually but steadily changing its technology base. "During the first transition, it was important to keep the company together while we moved all kinds of resources — not just money but people, infrastructure, expertise," he says. "In many ways, we're doing the same things today. Some of us have talked about how similar 1993 is to 1983. And so in many ways, we're executing the same kinds of strategies now: let's keep the company together enough so we can move people fairly painlessly, and make that the real investment strategy, rather than just sort of view it like a start-up."

> Past experience and solid management are the pillars on which renewal of firms is based.

Currently, PowerHouse is in use in 68 countries at more than 23,000 mid-range server installations. More than 15,000 copies of PowerPlay, Cognos's desktop product, are in use.

Setbacks and Mistakes

Although Cognos has been largely successful since it started, it hasn't escaped its share of setbacks — most of which occurred due to the rapid growth curve the company climbed in the mid-80s. "There have been a few setbacks," admits Minns. "We've invested too aggressively, forecasted more growth than we've been able to accommodate — that's arguably what led to some of the layoff activity. We were forecasting a certain amount of revenue, we were building a sales infrastructure and an R&D infrastructure to support that level of material, and it wasn't there."

> Balancing sales forecasts with the capability to support these sales is an imperfect science.

Two of Cognos's worst years were when it was under a so-called "professional" management style. After it scrapped both the manager and his style, the company reported eight consecutive quarters of profit. "We definitely, consciously made that decision [to move from entrepreneurial to professional management] when we hired Tom," says Potter, who staunchly defends Csathy's performance, emphasizing that he was exactly what the company needed — at the time. "But I think I have a slightly more balanced view today. I think that at great peril does any company in our business — at any size, by the way — compromise its passion for the technology and the market place in order to have good administration. We did suffer from that. And this isn't a comment on any one individual — everybody has their strengths and weaknesses. But our conscious decision to say, 'Let's get our administrative house in order,' compromised the passion for the technology, which has to come from the top.

"Now we're more than three times the size we were when Tom joined us," he adds. "But we've actually preserved what he brought in terms of good administration, and gone back to our roots and are driven by a passion for the technology."

Risks and Choices

"If you see that a product is getting a little long in the tooth, what do you do with it?" asks Minns, summing up the essence of the innovation dilemma. "Do you attempt to evolve it or do you attempt to replace it? We've sort of oscillated between these choices. The instinct of R&D people is always to have a new product. The instinct of most marketing people is to evolve what we already have and listen to what the customers are saying. It's hard, sometimes, to reconcile those two things."

> Marketing and R&D people have to work closely together to develop products that the market wants. This is easier said than done.

In a different area, one choice Potter made as soon as he took over the company was to make sure it was perceived as North American, rather than Canadian. "All our competitors in the 1980s — less so now — were American. That's why in 1987, as soon as we could, we began publicly trading in the US," he explains. "I think we felt it was really important to just ignore the US-Canada border. In so doing, we've become no less a Canadian company. One hundred percent of our R&D is still done here. We employ half our staff in Canada.

"You can be totally nationalistic, but it's still important to appear to be a North American firm," he continues. "It has to do with financing the company. The US is where the smart money is; it's more specialized, and there are more people involved. Nobody can be an institutional investor in Canada and specialize in the software tools business. You wouldn't have enough to do — whereas most of our investors are institutional in the US. They're shrewd, they're smart. That's an essential ingredient of entrepreneurship: to have very mobile capital, smart money that moves quickly to support your story."

It has nothing to do with aligning yourself with American values, he adds; it's purely a matter of survival and financial smarts. "I mean, do I like the short-term perspective of Wall Street? No. Do I believe in the social values of the US versus Canada? I don't. In other words, do you have to be less Canadian to compete? No. But you do have to access the best capital markets in the world, and that means the United States."

> To survive, Canadian technology firms need to have access to a more sophisticated US financial market. US investors have much more experience with technology firms.

Potter becomes defensive when asked if he considers Cognos's management style to have been comparatively conservative in the past. "I can tell you the aggressive growth across Canada that drove us in the mid-70s was not conservative," he says. "I can tell you that at the end of that decade we were stretched very thin as a company because we had expanded very quickly. I can recall seeing the scars of not being conservative enough because we were stretched thin financially, had a far-flung operation and were struggling to keep control over it."

He says the most important decision he ever made was to make the transition from the older business into the tools business. "You may say, 'Well sure, if something's taking off, go for it,'" he says. "Well, when we made that decision, 85 percent of our revenues were still coming from the consulting business, and only 15 percent from the products business. We sat down and said not only that we would invest more on products, but that we were actually going to get out of the 85 percent business.

"I think that was a very ballsy decision. I don't think there was anything conservative about that. It took two years to do it, and it was a huge challenge."

He concedes that Cognos is more conservative now, but thinks that's the way it should be. "I think that in a $150 million, publicly traded company, you have to be," he says. "You still have to be entrepreneurial. We organize to try and keep a lot of that spirit. But you have to answer to a lot of people, and you have to have predictable results."

The Outlook for the Future

The long-range prognosis for PowerHouse is something Potter and Zambonini have been considering for several years now, and it has resulted in a significant number of product releases in the last year and a half or so, based on Potter's conviction that centralized corporate applications will be increasingly extended to a variety of end users. Two new products, Impromptu and PowerPlay, both allow users to access, analyze and report on corporate data.

Desktop products are something Cognos will increasingly move into. They currently represent about 25 percent of sales, but the market is growing rapidly. "We are hoping, planning, investing to insure that that growth remains at 100 percent per year, at least for the next few years," says Potter. Cognos expects to double its revenue from desktop products in the next year or so. Eventually, that market may eclipse server-based products that have been the cornerstone of Cognos for most of its history.

"It's really clear to us that the shift towards client/server, which incorporates network environments, and PC-based interfaces, is a market that's just in its infancy," says Potter. "All of our R&D that's looking out more than a year or two is focused on those environments. Since we're just in the early stages, I think in the next five years we'll be reinventing ourselves as a client/server company. That's going to occupy a lot of our attention."

His view is that a fundamental shift will take Cognos's initial lead in end-user tools and make it a major feature of the company. He wants to maintain a balance between serving the professional developer programmer and the end-user. "By the end of the decade I see those things being almost indistinguishable," he says. "In the meantime, we'll have a balance, but we'll distinguish ourselves as having moved more toward the end-user side than almost any of our traditional competitors."

The Competition

One edge Cognos has over its competitors, says Minns, is its understanding of customers' needs. "That's probably because we were our own customer in the beginning," he says. "The software we have sold most successfully is software we invented for our own needs, and then generalized and sold more widely."

Cognos's main competitors in development tools, says Potter, are companies like Progress, Uniface and database companies such as Oracle and Sybase. "We believe the tools are clearly emerging as being independent of the database, so people can mix and match, take the best of both breeds," he says. "In the 1980s the strongest tools players were database companies who added tools to their database. They're still competitors today. Most of those competitors are sizable firms — some larger, some smaller, but most not less than $100 million." In end-user tools, the key competitors tend to be younger and smaller — $10 to $15 million companies.

> Many technology companies fail because they do not respond to customer needs. They believe that their technology is so good as to be irresistible.

"Increasingly, we have broadened our products and broadened our solutions to cover what the market still perceives to be two different areas of the business. So we tend to have two different sets of competitors," says Potter. "Our view is that we're ahead of most of those firms in understanding that there's ultimately one market. I don't like to bifurcate the market in terms of the way we view it, but the competitor set is different. One side of the market is tools for the professional developer, which is actually the market we entered in the early 1980s. The other is tools for the end user: desktop, data access, analysis. The reason we believe these two are actually the same market is that what we're really seeing is a steady transition from these kinds of applications being done by professional data processing people towards end-user computing."

Management Strategy

Never underestimate the importance of the international market, says Potter. "When we entered the products business, right from day one we targeted all the major international markets. Today it's even more important — and more challenging, because initially the markets that were most important were culturally more compatible. Now there are major opportunities in Japan, China and Latin America, and you're really dealing with some challenges. At a technical level, supporting Asian languages is an extremely difficult technical problem, very demanding, and very costly. But you have to be there."

> Going international increasingly means penetrating the fast-growing markets of South America and Asia, which requires cultural adaptation.

Regular management changes have played a large role in Potter's success formula, and he says he can't figure out why so many people ask him about these changes as though they indicate problems. "At the risk of sounding defensive, I think that a sign of danger in a company is if management isn't turning over at a faster rate than the technology is changing," he says, paraphrasing a favourite section of Computer Wars. "I think that's a bit extreme. But I find it ironic that I'm often called upon to explain why we've had [these changes]. I know that sounds defensive. But what I'm trying to say is we've had some tremendous value from some of the changes we've made — different people have done great things for us."

Although Tom Csathy is gone, the legacy of his good judgment lives on in the form of the managers he hired who are still around to exert a positive influence and provide guidance. The current chief financial officer, corporate controller and head of administrative organization, for instance, are all topnotch people Csathy brought to the company before he left. Says Potter, "We've had an enormous benefit from what Tom did. One of the great things he did for us was to implement financial controls and an administrative structure that still serve us today, even though we're twice the size we were when he left — and more than twice the size of when he arrived."

But, says Potter, by 1989 he had perceived a need for much more aggressive, motivational management, marketing and sales. "So we made a change and brought in a guy who really was masterful at that: Jeff Papows. What he did, particularly in the first year, absolutely galvanized the sales organization. Even though we didn't change anything about products or technology or markets, our productivity and sales performance went way up. And you could see why — he was just the toughest, most motivational, aggressive sales manager I've ever seen."

Today, he says, Cognos is addressing an entirely different set of needs. "We're in a period where the whole industry is shifting to a different technology base, and that involves technology, products, and markets. The guy managing our company today, Ron Zambonini, is, I think, the best in the industry in terms of having a vision that is steeped in the technology — a vision and an ability to communicate it.

"I don't want to suggest that I'm on some kind of a three-year cycle," he adds. "But I can certainly say that Tom Csathy and Jeff Papows gave great value to this company at the time when it was needed."

With almost a quarter century's worth of management expertise behind him, Potter has more than a few words of wisdom to impart. First, he says, it's absolutely vital to understand that technology is a people business. "I think most folks in the technology business understand that," he says. "But outsiders often assume [the whole game revolves around] one brilliant inventor — we're encouraged to view it as a personality game. But the knowledge base of the company is the only real asset. That's a long way of saying my first advice is to identify the key people in your organization and involve them as much as possible in the ownership of that company, as early as possible," he says.

> People are a technology firm's prime asset. To keep key people motivated, firms often introduce stock option and share purchase plans so that these individuals can have a sense of ownership and see their efforts rewarded.

"You can give people a sense of ownership in several ways — in the way you treat them, or by giving them a sense of ownership to ideas, to concepts. We also used, from a very early stage, stock options and employee share purchase plans, long before we became a public company. We had a lot of enthusiasm from our staff at that time, being part owners in the company. When the company then sees a high growth rate and its stock is valued at a very high level, small amounts become little nest eggs for people. I think we benefited from that when the strains of high growth hit us."

Potter also has a few words of encouragement to offer to anyone just starting out. "Because more mature markets in the technology sector tend to be consolidated, people might have a sense that opportunities for start-up are dwindling — that this is an industry where scale counts," he says. "I don't believe that at all. First of all, technology really thrives on fresh ideas that come not just from bright people, but from people who don't have a lot of baggage. It you've already got an installed base, you can't move as fast. Large organizations consider partnerships with independent companies to be part of their business.

"We do this. We're convinced that the services capability that our customers need can't come from us entirely, and so we form partnerships and find ways of encouraging people to become experts in our products and then offer those services to our customers. We compete with them. Why would you want to encourage someone who's a competitor? Because we can't do it all, and with their help, we're a stronger company, and customers are happier. In that kind of environment, there's a lot of room for small companies to start up. It's well established that the model of the computer business today is increasingly one of partnerships and inter-relationships between companies that don't actually have interlocking ownership.

"For all those reasons," he concludes, "it's a better time than ever to be an entrepreneur in this business."

14

Mitel Corp.: Renewal after Rapid Growth and Turbulence

The entrepreneurial spirit that drove the growth of Mitel was also the source of its difficulties when the founders 'bet the firm' on a new product that they could not deliver on time. Eventually new owners brought in the professional management needed to get the firm onto the next S curve.

If there was ever a blatant example of how destructive it can be for a young company to expand too rapidly, Mitel Corporation is it. A telecommunications business, it coasted from wild early success that lasted almost a decade to sudden near-ruin shortly afterwards. It led the high-tech pack straight up the mountain without so much as a glance over its shoulder — and when it reached the top, it toppled unceremoniously off. Mitel lived to tell the story, but it's now labouring slowly back up the mountain behind a series of more stable competitors.

In 1986, the year it fell off the top, Mitel was saved by British Telecom (BT), a UK-based telecommunications company, which stepped in and bought 51 percent of the company for $322 million. But Mitel's troubled years didn't end there. By 1990, BT had decided it no longer wanted its interest in Mitel, and publicly announced it was looking for a buyer. Two embarrassing years went by before the sale was finally made in June 1992: Schroder Ventures, the international management buy-out and venture capital affiliate of Schroders PLC, took Mitel off BT's hands for a meagre $60 million, or $1.40 a share.

Then the company's sagging fortunes took a sudden turn for the better. In what Rob Dietrich — an accountant, and vice-president of corporate affairs at Mitel — calls "the investment story of the year,"

Mitel's shares were sitting at five times that price by December 1993, or almost the same price BT originally paid.

Michael Cowpland and Terrence Matthews incorporated Mitel in 1971; full operations were underway by 1973. The two met when both were working as electrical engineers at Microsystems International Ltd., a subsidiary of Northern Telecom. In their late twenties and impatient with what they perceived to be Microsystems's conservative management style, they decided to quit their jobs and go into business for themselves. (Microsystems folded in 1975, after only six years in business, but the people who left, either before or after the firm failed, spawned more than 20 companies. The $30 million federal government investment in MIL was returned multifold through taxes.)

Cowpland and Matthews started with about $20,000 in combined savings. Then they convinced a group of Ottawa lawyers to back them, which was good for a further $100,000. Next, they drummed up some early business by knocking on a few Scandinavian doors (where Matthews had contacts) for their first large contract to design touchtone receivers. The contract was with International Telephone and Telegraph Corp. of New York. By the end of 1973, Mitel had fully launched itself on the telecommunications world.

> The origins of Mitel represent classic entreproneurial verve: from getting money wherever possible to stretching the size and importance of the firm.

In the beginning, everything the pair touched turned to proverbial gold. They seemed to have a magic combination of technical and strategic brilliance, energy, chemistry, determination and sheer nerve. They made the company seem bigger than it really was. They convinced potential customers that it possessed the importance to which it was pretending. On one occasion, when the company was still a four-man operation, an important potential customer asked to drop in and have a look around. Cowpland and Matthews tore around their leased office building, tacking Mitel signs up on other tenants' doors and convincing them to pose as Mitel employees. Most were doctors of sorts, and in their white lab coats made plausible quality control inspectors or technicians.

Sales doubled every year in each of the first five years of Mitel's operation, with the exception of 1979. After only six years in existence, Mitel was already standing out from the crowd because of its unique semi-conductors — silicon chips imprinted with hundreds of electric circuits, for use in all sorts of technologies. What was considered really exceptional was that Mitel designed its own semi-conductors, a fact that earned it a strong competitive advantage. Newspapers and trade publications seemed to be stuck on one favourite phrase to describe Mitel, referring to it again and again as the "darling" of the high-tech industry. Cowpland and Matthews themselves were alternately dubbed the "gold dust twins," or the "Kanata Dreamers." By 1982, Mitel had become one of the world's five fastest growing companies, with over $200 million in sales and 4,000 employees.

> Academia is an undervalued reservoir of commercializable ideas. The quickest route to the market place is for the researcher to run with his research results, as was the case with Cowpland.

"When we first started Mitel we had two great ideas we thought were surefire winners," says Cowpland. "It was Matthews and myself, and within the first month the ideas had totally failed. So we had to come up with a new idea. The third idea was based on my PhD thesis, which turned out to be the successful one. It's like third time lucky."

One of Mitel's first products was a device to allow companies to offer touchtone phones without having to entirely replace existing lines designed for rotary phones. This attracted the attention of some major American telephone companies, who still hadn't quite turned their heads south again when Mitel came out with its new private branch exchange (PBX) switchboard systems. Northern Telecom was ahead of Mitel with its own PBX system, but Mitel's version was cheaper and better, mostly due to a special chip they'd designed. Deregulation of the telephone industry in both Canada and the United States opened the floodgates for businesses such as Mitel; phone companies were prime targets for such innovative technology. Schools, hotels, hospitals and corporations of all sorts were perfect candidates, and newly freed telephone companies made grateful distributors. The number of employees at Mitel zoomed from 30 in 1975 to a record 6,140 in 1984.

But things started to crack around the edges a little in 1983. By 1985, they had caved in completely. Between 1983 and 1988, Mitel posted five consecutive years of substantial losses. The company essentially did itself in with rapid growth that it had neither the capital nor the management experience to sustain.

Now, with its shares up again, revenues climbing, losses a thing of the past, and new management to run things, Mitel is back on its feet, employing some 3,600 people worldwide and boasting a cumulative total of more than 168,000 sales of PBX systems in 80 countries.

History and Milestones

Mitel's first big success was the Touch Tone to Dial Pulse Converter. As its name implies, it allowed central switching offices designed for rotary phones to convert to touchtone devices instead. By 1977, Mitel had sold almost 50,000 of the converters, and was employing at least 200 people directly (not counting subcontracts). It had offices in more than a dozen countries around the world, and its founders were calling for $12 million in sales by the end of that fiscal year — and $100 million in five more. Its new product for 1977 was the Electronic Private Automatic Branch Exchange, known as EPABX.

By 1979, Mitel was emphasizing its new semi-conductors. The big news that year was a new production technique that allowed a family of microprocessors to operate five times faster than the industry standard, while consuming only one-fifth the energy. That year, Mitel set up a "think-tank" group of eight in Lake Tahoe, California, attempting to tap the vast pool of talent residing in nearby Silicon Valley. The group's mission was to generate innovations that would help the company grow to semi-conductor sales in the range of $90 million by the mid-1980s.

A technological lead spurs growth.

Mitel was at the head of the pack in that niche already, boasting that it could cram 30,000 transistors on a single chip, but it wanted to double that number within a year and keep moving after that. They were already producing a wide range of telecommunications products; one of the top sellers was the SX200 Superswitch, a computerized switchboard that could accommodate up to 200 lines. Those cost $11,000 each, and Mitel was shipping 100 of them every month. Plans to go public were already in the works. Exports to Europe and the United States were accounting for 60 to 70 percent of sales, and had doubled every year since the company started.

Around 1980, Mitel was awarded a $21 million federal government grant from Industry, Science and Technology Canada (ISTC), and used it to expand operations in Bromont, Quebec. The grant included a provision that Cowpland and Matthews allow the government first refusal on any sale of their shares to foreigners, including any sale that would reduce either of their individual holdings to less than 15 percent of Mitel's total outstanding shares. The agreement was to expire in March 1983. At the end of that year, Mitel announced a $20 million expansion of its Kanata facility, to allow space for two new products and the 1,000 extra jobs they would create. The top new product was the SX2000. Enabling the connection of up to

10,000 lines, it was the most concentrated chip in the world. The other PBX system, called the Super 10 system, was designed for small businesses to handle 16 extensions and 8 outside lines.

The big coup for 1981 was that Mitel was unexpectedly permitted to set up a wholly owned subsidiary in France. Up until this point, the French government had insisted that all foreign-owned electronics companies sell or license their technologies to locally owned companies, or arrange joint ventures. This year, Mitel was still predicting 1982 sales of more than $220 million, and $1 billion by 1985.

In 1982, Mitel established a Japanese subsidiary and opened a sales office in Tokyo. The federal government kicked in $9.9 million in grant money from the Enterprise Development Program, which Mitel earmarked to further develop its hot new product, the SX2000 private automatic branch exchange, designed to accommodate 150 to 10,000 lines. Part of the company's strategy was that each order for an SX2000 would bring in great numbers of orders for associated products, such as terminals and add-ons that could be plugged into the switch. A new digital product, the exchange was small, powerful, and easily adaptable to European standards — and largely expected to help Mitel invade the Fortune 1000 market. This was a lofty ambition for a company not quite a decade old, as it put Mitel in direct competition with Northern Telecom — already a giant in digital communications — as well as with other big American computer and telecommunication companies. The industry consensus was that the SX2000 would either make or break Mitel. Cowpland and Matthews didn't know it yet, but it would also end up devouring about $100 million in research and development costs.

Early in 1982, Mitel encountered a major setback when the American government put an end to AT&T's monopoly in the United States, disabling Mitel's best customer. The decision caught Mitel unprepared, and it lost a quarter of its market within two months.

About halfway through 1982, rumours began to circulate about the possibility that Mitel would be supplying IBM with the SX2000, which the two companies had been developing together. In 1983, Mitel's incredible good fortune came to a screeching halt. Profits were down by half; revenues increased only marginally. What made the year so important was that its results would shape the years to follow: Mitel was still predicting sales of $1 billion by the end of the decade, and this goal had been shaping the company's management strategies. Late in 1983, the company reorganized its senior executives, with Cowpland and Matthews swapping jobs. Cowpland stepped down as president and became chairman, and Matthews became president and CEO; Cowpland would continue to guide technical developments.

Shortly after that, an international management consulting firm was called in to improve Mitel's organizational effectiveness in preparation for the following 10 years, when executives anticipated Mitel would become a $1 billion company. By now, the SX2000 was seriously behind schedule, and bigger companies were starting to crowd the market. Firms like Nippon Electric in Japan and AT&T in New York were starting to stake claims to a market that might have belonged to Mitel had the SX2000 made it there on schedule. One of the first indications that something was very, very wrong at Mitel was the closure, at the end of 1982, of a brand new, $10 million, 50,000-square-foot plant whose construction was just nearing completion in Buctouche, New Brunswick.

By spring 1983, customers, distributors, investors and financial analysts were becoming uneasy about the delayed delivery of the SX2000, which had been scheduled for late 1982 or early 1983. This year would be the end of Mitel's phase of runaway success and exponential profits. Fans and critics both still agreed that if Mitel could get the product out, it would revolutionize the market. But now they were adding that Mitel had "really gone out on a limb" this time, and investors were edgily watching competitors catch up. It was generally assumed that a delay until fall 1983 would be acceptable, but anything later than that would certainly jeopardize Mitel's credibility and prospects. By June, with no sign yet of the SX2000, IBM dropped an anticipated deal with Mitel, electing instead to do business with one of Mitel's biggest rivals, Rolm Corp., which it also bought. This was viewed by most as an enormous setback.

> Delays in bringing a promised product to the market can seriously undermine the credibility of a firm.

Fiscal 1984 was Mitel's first year of losses in its 10-year history; it reported losses of $32.4 million. Losses plagued the company throughout most of fiscal 1985, totalling $32.1 million by the end of the year — almost the same as the previous year's total. Intense competition from American telecommunications companies showed no sign of letting up. Mitel's share of the PBX market dropped by just over a percentage point from the previous year, from 11 to 9.9 percent, while the overall market grew by 23 percent. The company blamed the losses on expensive production and start-up costs for the SX2000, and on lower sales and profit margins caused by heavy competition from the United States. No one was sure, by this time, whether or not Mitel had either the financial or the marketing resources to continue to survive in this market. That year, Mitel struck a $44 million deal to supply telecommunications products for a year to BT.

Early in 1985, speculation began concerning the possibility of Mitel being bought out by British Telecom. By September of that year, it was almost a certainty: British Telecommunications PLC would purchase 51 percent of Mitel for $322 million. Shortly after that announcement, Mitel laid off 440 employees. By the time the deal was formally approved in February 1986, Mitel had posted losses in 9 of the previous 11 quarters, for a total of $110 million in losses. A month later, Cowpland gave up his position as chairman to Matthews, maintaining only a seat on the board and some influence on technology development. This was the first time since 1973 that Cowpland was not involved in the daily operations of Mitel. He said he would use the time to pursue other business interests. (As it turned out, he used it to found Corel Corp., a successful graphics software company discussed in Chapter 10.)

> A change of ownership usually brings with it structural and managerial consequences.

Matthews replaced Cowpland as chairman of Mitel, while British Telecom found it a new president: Anthony Griffiths. Griffiths had started with a BA from McGill in 1954, gone straight to Harvard to do an MBA, graduated in 1956, and then launched his career. When Mitel found him, he was a partner at Connor Clark and Co. Ltd., an investment management company. Griffiths's mission would be to turn Mitel back into the runaway success it once had been. At the time of his appointment, Mitel had suffered losses in seven of the previous nine quarters.

Griffiths's first move was to restructure and centralize management of Mitel's regional units. His intent was to streamline the company's structure, simplify decision-making processes and return the company to profitability with all possible speed. In 1986, after losses of $160 million in that fiscal year, Griffiths warned that it might take two or three years to accomplish this.

In 1987, Griffiths became chairman, appointing a new president and CEO, John Jarvis, who immediately laid off 400 people in addition to the hundreds Griffiths had already laid off. By 1988, Mitel had been reporting steady losses for five consecutive years. Things had improved slightly when BT bought 51 percent of the company, wiping out its $260 million debt; but even after that, Mitel continued to report losses. Its stock, meanwhile, had plummeted from a high of $37.25 in 1983 to a low of $2.70 in 1988. During this year, Mitel sold off a subsidiary in Toronto and laid off about 40 more people in Florida.

In April 1988, financial analysts were figuring Mitel would break even by 1989. In 1989, Mitel defined its mission as identifying and then filling new market niches — for instance, systems for specific industries such as hospitality and medicine. By now, seven of 10 senior managers had only been with the company since the summer of 1987.

> The realization that a major product line has reached the top of the S curve sets in train a search for new directions.

By 1990, Jarvis and Griffiths were saying the market for PBX products had leveled off, with prices declining. Early in the year, British Telecom announced plans to sell its 51 percent share in the company, but there were no takers. BT asked Griffiths to take over as interim president until a buyer could be found. For the second time in a year, Mitel asked employees to help the firm cut costs and avoid layoffs by resigning or taking extended leaves of absence.

Mitel began designing its VX product family — voice-processing products designed to operate with both the SX200 analog system and the SX50 private branch exchange. The idea was to create a series of products with a core technology that could be exploited other ways. An example is a voice-messaging system designed for hotels — it saves staff taking messages, and benefits guests by letting them receive detailed messages privately and in their own languages. The idea was to expand this sort of operation to hospitals. Mitel had, by now, identified two of its major markets as lodging and health care. Research on this technology continued throughout 1991.

Part of 1990's problems can be blamed on a declining market. The total PBX market declined by 7 percent in the US, by 13 percent in Canada and by 16 percent in the UK. The positive side to these figures is that the declines were faced evenly by all of Mitel's competitors. Mitel held onto its market share.

In April 1991, Mitel announced layoffs of 10 percent of its workforce — another 400 employees — and closed its US sales office in Boca Raton, Florida, in anticipation of major losses for the 1991 fiscal year. Losses — which ended up totalling $107 million — were chalked up to the leveling off of the PBX market and the worldwide recession, as well as to overcapacity, duplicated efforts and expenses that exceeded the company's level of operations.

By early 1992, BT was still waiting for a buyer to take 51 percent of Mitel off its hands; the board of directors declared the "resolution" of the matter to be "among the highest priorities." The company laid off close to 1,000 people that year, saving $40 million. American and Canadian sales operations were consolidated. New Zealand distribution was transferred to Telecom Corporation of New Zealand. Mitel acquired Telenova Distribution Company Inc., headquartered in New Jersey, for $13.5 million, renaming it Mitel Telephone Systems Inc.

The demand for PBX systems continued to decline that year — down by 3 percent in Canada, 7 percent in the United States, and almost 11 percent in the UK. Losses went from $107 million in 1991 to $5.7 million the following year, and back up again to $12.6 million by halfway through fiscal 1993.

"We had some write-offs in some of those years," says Dietrich, explaining the losses. "Most recently the losses stem from recessionary problems compounded by the period of uncertainty when BT announced they were selling. It took them two and a half years [to find a buyer], when we had this cloud hanging over our heads, not knowing whether we were really going to be in business, or whether someone would buy BT's share, and I don't think that helped our sales at all. It was very difficult because competitors would ask potential customers whether they really wanted to buy systems from a company that BT didn't want any more. It was a very, very difficult time for us. There were a number of reasons for the losses, but they all boiled down to not enough sales."

> A new owner can create the stability to get the firm onto the next S curve.

Since then, says Dietrich, "There's really been a change." Schroder's purchase of the company stabilized things, and during the time Mitel was waiting to see who would finally buy it, the company had continued to spend heavily on R&D. "The new ownership and refreshed product line hit the street at the same time, and there was a complete reversal," says Dietrich. "We're on a roll now [December 1993]. People have a lot of confidence in us, our stock is up, people think we must be a great company, so they buy our product — you get the whole opposite reaction."

Mitel's current president and CEO is Dr. John Millard, formerly a senior vice-president at NEC America. Griffiths finally left Mitel early in 1993 and went to Toronto to pursue his other interests. "Tony is on the board of directors of a number of Canadian companies," says Dietrich. "He does a lot of investing and consulting. When he pursues his own interests, he's got a lot of things to pursue."

As a company with just under $500 million in sales annually, Mitel doesn't rely on government funding. But in its earlier years it received money from several different programs: $383,000 in 1980 from the Department of Regional Economic Expansion; $1.7 million from the Defence Industry Productivity Program that same year; and $20.9 million from ISTC, spread over five years from 1979 to 1983.

Spending on R&D will amount to 9 or 10 percent of revenues this year, or about $50 million. "It's historically been a bit higher than that, but revenues are going up dramatically this year, and R&D spending is relatively flat," Dietrich says.

The company has 3,600 employees, 1,400 of whom work in Kanata. About 200 more work elsewhere in Canada, and the remainder are distributed around the world.

Dietrich says about 75 percent of Mitel's total revenues come from its smaller PBX products, which are sold in three main geographical areas: North America, Europe (primarily the UK) and the Pacific Rim — Hong Kong, China, and Thailand. About 15 percent of the business is in worldwide semi-conductor sales. Only 10 percent of total sales are in Canada, and the rest are around the world — Brazil, China, the US, Japan. "We have a broad range of products," says Dietrich. "The basic point of the construction of Mitel today is that the largest proportion of R&D is done in Canada, and the majority of our sales are gleaned from outside of Canada. In the PBX business, the bulk of the end users are generally companies with 50 or more employees.

"We have some very large customers, such as the Metro Toronto Police Force, schoolboards and government agencies," he says, "but Mitel has made its name in selling small PBXs, and we've sold probably over 170,000 in the history of the company. The average customer at Mitel is probably a small business somewhere that you never heard of."

The role of telephone companies has changed dramatically since the early 1980s, when they made excellent distributors for Mitel and were therefore important channels to the market. "Now," says Dietrich, "we sell our own products directly to those customers with our own sales force, or through other distributors. In Canada today, the telephone companies are still probably the most important channel for our product, but nowhere else is that the case."

Mitel's most important product now, says Dietrich, is the new LIGHT series of PBXs, introduced in 1993. There are two products, one derived from the SX200 product line and the other from the

SX2000. LIGHT products are connected with a single strand of optical fibre, designed to operate in what Mitel calls "campus-like" business and government environments, over long distances.

The SX200 LIGHT was introduced first, and so far, says Dietrich, the response has been outstanding. "Sales have exceeded our expectations for both of those products," he says. "At the same time, we also introduced new telephone sets that work with these products."

Sales of these new products account for a large part of the dramatic increase in revenues. "We also have much higher revenues in PBXs overall, and sales of electronics are the bulk of the PBX sales," says Dietrich. "There's been a resurgence in the PBX market in the US, and we've gained some market share this past year. We think our new products have had a lot to do with that."

The newest product is dubbed Radicall. It's a product line that includes a compact, wall-mounted device that enables telephone companies or other network providers to deliver their services and features to the desktop using Mitel's Superset 400 series of telephones. Radicall incorporates Mitel's spreadsheet software package, allowing the network service provider to create customized features and service applications for both workgroup environments and individual users.

Mitel has identified its priority for 1994 as more aggressive marketing, despite the corresponding increase in cost. The goal is to have an expanded customer base across a wider geographic area.

Setbacks and Mistakes

A setback in the mid-80s nearly extinguished Mitel. The delay of the SX2000 private automatic branch exchange snowballed into numerous other problems with image, reputation and capital. Mitel began to report heavy losses. The company's executive vice-president, Donald Gibbs, left. Cowpland sold about 700,000 of his shares, reducing his stake in the company to about 13 percent. About 400 employees in the US were laid off. The semi-conductor manufacturing plant in South Burlington, Vermont, was closed down. The company's Irish subsidiary closed. On a single positive note, by the end of 1986, the SX-2000 was finally on the market, enjoying brisk sales, and Mitel had plans to introduce new products in the upcoming fiscal year.

> Betting on a new major product and not being able to deliver sets up a domino effect with very serious consequences.

One of Mitel's greatest problems, according to analysts, was that its excellent products and rapid innovation caused swift growth that was not matched by expertise in sales, marketing or management. The criticism is that Mitel's early products had sold themselves, so no one had to learn how. Also, its streamlined production plans and concentration on the SX2000 left much of its inventory obsolete. Sales lagged behind projections, and the company expanded too rapidly.

"We had higher expectations than we were able to realize," says Dietrich. "I think in the mid-80s we had a view that this company was on a growth path that would take it to $1 billion within a certain time frame. As a result, the worry was whether or not the company would be able to manufacture the product to take it to a billion. The reaction to that concern was, 'Let's build some manufacturing plants so we'll be able to do the manufacturing when the orders come in.'

"The legacy of that," says Dietrich, "has taken us a number of years to resolve. We've closed plants all over the world in the last seven years and boiled down our manufacturing facilities here to three or four, while we had in excess of ten in the early 1980s. That severely hampered the company through the back half of the 1980s, basically because we were dealing with too high a cost structure.

"But," he adds, "at the end of the day it hasn't been bad from a competitive point of view, because we had to downsize ahead of these giant companies that are going through it now. Other companies are reeling under their plant capacities and too many employees, whereas we've gone through all of that already. We're on an upswing while some of them are on a downswing, and that's to our advantage."

A more recent problem was the declining markets for PBX equipment, which in 1992 made up the bulk of Mitel's business. In 1993, things finally began to pick up. "The PBX market had been declining at a rate of about 5 percent per annum in the US, and that was quite worrisome," says Dietrich. "But it's turned around just this year [1993] — and in the first half of this year the market was growing. Certainly the market is growing if you look at the Far East, because of the rapid growth of the economy over there. There's still a slight decline in Europe."

Dietrich believes the driving force behind PBX sales is economic growth. "So in the recessionary period we've had in North America in the last few years, it hasn't exactly been boom times for selling telephone systems," he says. "When somebody lays off 20,000 people, their first thought isn't to rush out and buy a new PBX system."

Mitel's only other weakness now, says Dietrich, is its small size, relative to its competitors. "On the other hand, our relative strength against bigger competitors is that we can move a lot more quickly than they can," he says. "We can be more innovative, we're not bogged down by the bureaucracy of a big company."

> Size is double-edged. A large company has the presence and staying power but can be attacked by smaller, more innovative companies, as was the case for IBM.

Risks and Choices

Differences in opinion about how to run the company started to surface shortly after the SX2000 fiasco. To some, it was unclear who was actually in charge. While business was going down the tubes at Mitel, Cowpland had been investing money in other emerging technologies. When these failed, he was forced to sell Mitel shares to cover his losses. Some analysts said the fact that one of the company's founders was selling shares caused Mitel's stock to plummet even further.

The basis for the disagreement over how best to manage the company was the choice between a conservative management style or full-speed-ahead, entrepreneurial pizzazz. Some top executives worried that Matthews was "betting the whole company" on the SX2000, and that Mitel did not have the financial resources to support it. The SX2000, they argued, was a bottomless pit that was consuming funds at a terrific rate. As a result, funds for the smaller products that had been Mitel's meat and potatoes dried up, and Mitel lost some of its competitive edge in an arena that had once been its home base. In retrospect, it looks as though Mitel's founders and executives were still being propelled by the adrenalin rush created by the company's whirlwind start. Instead of entrepreneurial spirit eventually giving way to more polished, professional management tactics, management at Mitel seemed to grow less careful and more risk-taking and informal by the day.

> Knowing when to shift from entrepreneurial to professional management is key.

Also, said analysts, Mitel bet all its chips on the SX2000 at exactly the same time that the semi-conductor market began to shrink. Many in the industry trace the start of the trouble back to an informal agreement Mitel had struck with IBM to supply the SX2000. It's only conjecture, but the consensus seems to be that the IBM agreement gave Mitel a false sense of its own size, sparking the swift expansion that the company found itself unable to support after IBM dropped out of the deal.

"My understanding, just from talking to people in the industry," says Dietrich, "was that IBM had a view that it had to align itself with a company that had PBX technology, because they also had a view of the integration of telephones and computers in the early 1980s. They were evaluating Mitel and a rival company, Rolm Corp., in the US. They ultimately selected Rolm as the company they would work with, as opposed to Mitel — much to the disappointment of people at Mitel who thought the deal with IBM was locked up. That's when British Telecom was identified as an investor in the company.

"I think it was perceived that the reason IBM chose not to proceed with Mitel was that there were problems with the SX2000 — which, in fact, materialized — and so therefore the company's fortunes faded," he says. "But perhaps if IBM had stayed we would have had the same problem — the product took some time to get to the market.

"To tell the end of the story," says Dietrich, "the SX2000 is our flagship product today. We built a Cadillac that's turned out to be a Cadillac. The only problem was the time it took to reach maturity."

Maturity has meant six consecutive profitable quarters ending in June of 1994. Profits rose dramatically from $2.6 million in fiscal 1993 to $20.7 million in 1994. Millard will use the cash to pursue international alliances and to increase spending on research and development.

The Outlook for the Future

Now that the market for PBX equipment appears to be declining every year, Mitel is looking for new niches among businesses that are busy downsizing, reorganizing and specializing — Mitel anticipates a need for newer, more flexible approaches to communications, and wants to be prepared to deal with the possibility that PBX systems may eventually be obsolete. To that end, they have been developing, for two years now, the SX200 LIGHT and the SX2000 LIGHT. These systems are designed to complement and support the new SUPERSET 400 series of business telephones.

"While PBX sales were dropping, sales of centrex — smaller size units — were growing," says Dietrich, explaining the company's new strategy for Radicall. "That's a market that's going to grow. It also looks like centrex will be introduced by telephone companies in Europe. But the area we really think is going to boom in the longer term is what we're calling 'computer-related telephony.' Basically what we think will happen is that the PBX may well disappear in the long term and be taken over by telephone functionality that resides either in a local area network or through the ATM technology that's going to be employed in the late 1990s.

"Our strategy is to provide telephone applications that will run in this computing environment. We're now developing and introducing products to make sure that when that happens, we still have some sales," says Dietrich.

The Competition

Mitel reported its first loss ever just one month after IBM backed out of the tentative deal to buy SX2000 systems from the company — Mitel wasn't big enough yet to support all the promises it was making about production schedules. When IBM grew impatient with the delays, it knew exactly where to look: it allied itself instead with Rolm Corp. of Santa Clara, California, one of Mitel's American rivals. After Griffiths and Jarvis took over, Mitel returned to specializing in the small-PBX market, where it is still a leader.

Now, Mitel has about half a dozen major competitors in the world in telecommunications: Alcatel (a French company), AT&T, Siemens, Northern Telecom, NEC, and Fujitsu, in order of size, and depending on which national market is concerned. "In the PBX business we're number five, maybe number six in the US," says Dietrich, "and maybe number six worldwide." In Canada Mitel is right behind Northern Telecom, with about 25 percent of the market. Mitel has about 7 percent of the market in the US, and 15 to 20 percent in the UK.

"In a flat or declining market, chasing after market share against the biggest companies in the world really isn't a brilliant strategy," Dietrich notes. "The whole point is that if we strike out in a bit of a tangential area, where we can take our expertise but where others aren't playing, we view our chances at a large market share of a new market as being much greater than trying to take two market share points away from Northern Telecom in the United States."

Management Strategy

Without a doubt, says Dietrich, Mitel's greatest strength has been the resilience of its employees. "We've gone through a tremendous amount of change in this company throughout our history in terms of the different presidents, owners and financial troubles that we've had," he explains. "And through all the turbulence we've managed to maintain a fairly high degree of enthusiasm, and continued to get new products out into the market, and I think that's directly attributable to the people we have working here." He also credits strong distribution channels, especially in the United States, and "world-leading" technology, where Mitel is known for its user interfaces in its telephone sets. "That's probably our greatest asset in the PBX area," says Dietrich. "I think one thing that really got Mitel going in the beginning was our ability to take the technology and package it in a way that was very easy to use at a low cost. That's how we really made our name — selling low-cost PBXs to small businesses. Large businesses were being looked after. But nobody else was offering that cost-value relationship down into where the mass market was. So we did that, and that's how we made our name in the early '80s."

> As always, people are the major asset.

He doubts Mitel will ever again see the kind of rapid growth it enjoyed in the early 1980s, but isn't sure that would be a good thing anyway. "That kind of exponential growth, doubling per year, I can't imagine it," he says. "It's really difficult to get that kind of growth. We probably wouldn't want to grow that fast." Steady growth in the area of 15 to 20 percent each year, says Dietrich, would be "a major accomplishment."

At just under $500 million in sales annually, Mitel will have to strike a balance between the entrepreneurial management that drove it in the early days and the bureaucracy that arrived with British Telecom in 1986. "There was a dramatic change in the business when British Telecom invested in it," says Dietrich. "Certain things they wanted to see in the management of the company were really different from what an entrepreneurial company would be — a much more conservative style, more operational controls, et cetera."

Was it ultimately good for Mitel?

"I think it would be fair to say that anyone who was associated with the company in the early '80s felt that as the company grew it needed more control, better management systems," says Dietrich. "But British Telecom is an institution as a telephone company, and a telephone company management asserted on top of a manufacturing concern, whether it was in or out of control, would be a burden not many people would enjoy. A telephone company is a much different operation than a manufacturing company. I mean, the whole goal of what we're doing is to bring product to market as fast as possible. In a telephone company — a bureaucracy — they try to slow things down. So you're really at odds."

In Dietrich's estimation, Mitel is on the bottom of its second growth curve. "We've gone through one already," he says. "We got to the point where the S started to slope down again — and then we managed to get back up on the ramp of the second one. What we've been doing for the last two years is moving ourselves back up on this next curve. Our results were almost flat for five years in a row until two years ago. Now we're back on a phase of 15 to 20 percent revenue growth again."

A question that remains unanswered is why it took British Telecom so long — two and a half years — to sell its stake in Mitel. The initial assumption, Dietrich says, was that a competitor would buy Mitel for market share. That didn't happen, probably because the company's market shares weren't outstanding, particularly in the United States. The implicit suggestion is that many competitors who looked at Mitel and thought of acquiring it wondered if so much trouble could justify such a small additional share of the market.

It's not an issue Dietrich likes to dwell on, and he hesitates to speculate about the possible explanations.

"That's a difficult question," he says. Then he adds, "It's history."

15

Gandalf Technologies Inc.: Rationalizing Too Many Products

Growing by rapidly introducing new products can be a successful strategy. But eventually the costs of maintaining a large array of distinct products can become very costly, forcing rationalization and a focused strategy. This is such a story.

Founded in 1970 by entrepreneurs Desmond Cunningham and Colin Patterson, Gandalf Technologies Inc. is now a $160 million Ottawa-based company that designs, manufactures and supplies a broad range of computer communication and information networking products, systems and devices. It employs some 1,400 people worldwide, almost half of them in Ottawa, with engineering, manufacturing, system assembly, testing and approval operations in Canada, the United States and the United Kingdom.

Gandalf is the name of the wizard with the magic touch in J.R.R. Tolkien's *Lord of the Rings* trilogy. It's no coincidence that the founders also chose this whimsical name for their company. Although there have been times when Gandalf waved its wand and found the magic was gone, its overall success is well-documented in the Ottawa high-tech community.

"The main market shift that created an opportunity for Gandalf was deregulation," says Wendy Burgess, the company's vice-president of quality and communications. "In the old days, if you wanted to send data or use a modem, you had to rent from a telephone company. Then the CRTC ruled that you could buy those devices; as well, you could buy phones from other companies. That really opened up a window of opportunity for companies like Gandalf to provide less expensive modems than what you would get from the telephone company. And you could buy them outright instead of leasing them."

Burgess has been with Gandalf since the early 1980s. She started off as an electronics technologist, moved into several other technical jobs, went into marketing, and then, she says, she was sent on what's known as the "Des Cunningham scholarship" — back to school for a year to earn a business degree.

Cunningham, whom Burgess refers to as "a local hero," has tried to retire several times, but keeps getting persuaded to come back whenever the company "gets into trouble," she says. Since the company began, Cunningham and Patterson have traded positions several times, taking turns at being president. From the start, it was a sound, productive partnership, says Burgess. "Colin is the technology person and Des is the marketing strategist. It was a good combination. One had the technology and the other knew how to sell it. Des is very practical in operations, too. He sees the long term as well as the details."

Gandalf's first product was a modem that offered users an alternative to existing, more complex, more expensive data transmission devices. PACX, Gandalf's first data switch, created a worldwide market in which Gandalf led the pack for nearly two decades, despite some setbacks midway and a rough ride through the early 1980s. Gandalf's innovation in data transmission allows customers to communicate with a variety of completely different computer systems, as well as with terminal and PC users in local and wide area environments.

> It is common for a founder to have difficulty in letting go of his/her company. At times the company has difficulty letting go of the founder.

Gandalf now organizes its functions into three different divisions. These are products that do multimedia networking in both local and wide area environments; multivendor network management systems designed to integrate Simple Network Management Protocol (SNMP) and Open Systems Interconnect (OSI) input; and products and systems designed to connect with the Integrated Services Digital Network (ISDN).

Gandalf scored a coup in its ongoing battle as the tough little scrapper in an arena of giants in its merger with Infotron Systems Corporation of Cherry Hill, New Jersey, in 1991. The merger, in which Gandalf acquired 100 percent of Infotron, allowed Gandalf to combine local area networking (LAN) with Infotron's wide area networking (WAN). The result was enterprise-wide networks, a much broader market, and growing profits.

History and Milestones

Gandalf's first key customer, says Burgess, was the federal government's Communications Research Centre, which bought a local data set — an inexpensive, short distance modem device. Cunningham is still grateful for their early support, says Burgess, because of their unexpected willingness to take a chance on something as risky as "a couple of guys making stuff in their basement." McGill University was another important early customer. Their first package was a data switch to allow students and teachers to switch from one computer to another.

In 1970 — situated in Manotick, Ontario, because that's where Cunningham lived — Gandalf Data Ltd. started to design and manufacture local and medium distance modems. The idea came out of Cunningham's and Patterson's recognition that businesses would increasingly want to link computers to other offices and central institutions. Throughout the 1970s, most of Gandalf's products were built on this initial idea. Over two decades, the first, simple, local modems evolved into ones that could internetwork with cities, countries and eventually continents.

By 1972, Cunningham and Patterson had begun to include the design and commercial manufacture of a data switch, the Gandalf PACX, setting a new standard and creating a global market. The first microprocessor-based statistical multiplexer was delivered in 1976. By the end of the decade, the company was spending about 10 percent of its revenues on research and development, and exporting its products aggressively worldwide. Patterson, president at the time, believed that research and development would be the basis of Gandalf's product line in the future, and that export was a matter of basic survival. The goal was to leave no market untapped.

By the beginning of the next decade, with its wagon hitched firmly to the rising star of the data communications industry, Gandalf grew rapidly. In 1980, the company made $26 million in sales, doubling the previous year's figure. That year marked the introduction of a new innovation on the multiplexer: a networking statistical multiplexer, introducing the concept of private multinode networking. By 1981, sales had jumped to $40 million, but the company was finding itself in a cash-flow squeeze because of all this rapid growth — they were expanding so fast that it was becoming difficult to generate

---

Government as the first "friendly buyer" of advanced products provides an important stimulus and credibility for start-ups. Governments in most industrialized countries use procurement as an instrument of industrial development policy.

enough revenue to keep up with the materials required for production. "They don't teach that in the textbooks," then-president Patterson said at the time.

To support its growth, the company decided to go public, with an offer of just over 2 million shares, raising a total of about $15 million. By 1982, Des Cunningham had replaced Patterson as president — Patterson would be vice-president of technology for the next four years — and was predicting future annual growth rates as high as 50 percent, although he foresaw that such growth might be hampered by the recession. That year, Gandalf introduced its Pacxnet data network, beginning a cycle that would considerably widen the markets in which it operated — and which would require broader and costlier development initiatives. This was also the year in which Gandalf scored a $1.5 million sale of its mobile taxi terminals to Houston for the Greater Houston Transportation Company's yellow cab fleet.

Selling to about 20 different countries, and trying to deepen existing markets rather than to create or enter new ones, the company reached record profit levels that year that have not since been repeated — $6.9 million, or 73 cents a share. By the end of the following fiscal year, 1983, those figures had dropped to $3.7 million and 38 cents a share; sales were up by a mere 9 percent from 1982. Cunningham attributed the drop in profits to four different factors: the recession; a shift to direct marketing in the United States and Britain, which increased marketing expenses by 21 percent of sales; increased competition in data communications, from both large and smaller start-up companies; and more vigorous research and development — expensive, but necessary in order to keep up with the competition. Also, the company's products were growing more and more complex, requiring added spending on technical field support. Gandalf ended up having to revamp almost the entire product line at once. Spending on R&D leapt from 7.9 percent of sales in 1982 to 11.1 percent in 1983 and up again to 13 percent in 1984.

The new innovation for 1983 was the Advanced Network Manager, a graphics-based network management system. In the following year, 1984, Gandalf began work on a fully distributed network data switch, the PACX 2000, designed to support 25,000 subscribers. Two people went to Australia to establish a sales base for the Southeast Asia market. The company opened a technology centre near London to promote business in Europe, particularly with British Telecom. It acquired a 99 percent share of Redifacts Advanced

190 The New Innovators

Manufacturing Aids Ltd. of Ottawa, a designer of factory-floor data-collection systems centred on the clothing industry. It was preparing to deliver software for its computerized taxi dispatch system, which by now was already installed in six Canadian cities and on order in three American ones for 1985. The company's strategy was firmly market-driven now; analysts were heralding Gandalf as one of the big turnaround success stories of the year because of its impressive comeback from slowed sales and profits in 1982.

By now, the intensive R&D efforts first implemented in 1982 were beginning to show results in the market place: 95 percent of Gandalf's 1984 sales were derived from products developed in the preceding 18 months, including various centralized data switching products, enhanced software, and a greater range of switch configurations. Also in 1984, the company acquired OCRA Communications Inc., renaming it Gandalf Systems Group. The subsidiary began testing systems in federal government offices and gearing up to compete in rapidly growing markets such as electronic mail.

The company's challenge now was to establish a more efficient management strategy — one that would provide some control and direction without squashing its researchers' enthusiasm for innovation. Three new vice-presidents were appointed to jointly run Gandalf's three previously separate divisions in the US, Canada and Britain; the three VPs would together oversee and coordinate marketing, manufacturing and technology. Although he was certain of the effectiveness of the new system in accounting, Cunningham admitted at the time that the company probably ought to be a little more freewheeling in different areas. The ideal was to reach a midway point between Gandalf's traditionally conservative approach and a more risky, entrepreneurial one.

Providing a management structure that gives direction while maintaining creativity is a constant challenge.

In 1985, Patterson became president again. Gandalf broke into the Toronto taxicab market, selling a computerized dispatch system to the Diamond Taxicab Association for $1.2 million. The contract required Gandalf to install the system in 400 Toronto cabs — the largest system in North America to date. By now, Gandalf had already sold similar systems in three other Canadian cities and installed them in some 1,200 American taxis. More systems were on order for cab companies in Anaheim, Indianapolis and New York City. The system could manage 800 cabs simultaneously. Each cab is equipped with a small visual display unit to relay messages to drivers; a small keypad is attached for the driver to use when sending messages back to the dispatch office.

In 1986, Cunningham became president again. "Des had been wanting to retire as CEO for some time," Burgess explains. "But we kept bringing him in when we got into trouble, to help save the day. There was a lot of him retiring, then coming back in to save the company." Gandalf's strategy for long-term growth was to sacrifice profits in exchange for advancing its market share. Sales still soared, but a hefty amount of cash was being reinvested in R&D: 13 percent of 1985's sales of about $85 million. Still, with only $5.4 million in long-term debt and equity of almost $50 million for fiscal 1985, analysts were calling Gandalf's balance sheet "a thing of beauty," wondering only when Gandalf's net margins would return to normal.

Gandalf's sales strengthened in both Europe and the United States that year. By now, Europe — including Britain — was generating about a third of Gandalf's revenues, with the US market responsible for almost 40 percent, and Canada and the Pacific Rim making up the remainder; Europe had emerged as the company's secondary market, replacing Canada. Despite strong sales, the company still had only a 10 percent world market share in the data transmission and networking business — and only a 3 percent share in the American market. The long-term goal was to move in the direction of making a complete networking system to enable different computers within an organization and its affiliates to communicate.

> Trading profits for market share is a well-known strategy. It contributed in large part to the success of Japanese firms.

In 1987, Cunningham stepped aside once more as president, embarking on a two-year retirement and making room for James Bailey, formerly the executive vice-president. That year, researchers at Gandalf came up with what some people called "the missing link": a hybrid networking system, known as the Starmaster intelligent network processor, that would connect all subscribers and resources in multivendor environments. Starmaster meant customers could establish communication among completely different systems, such as Apple, Digital and IBM (and clones). The product was heralded as a major breakthrough by those in the industry, and Gandalf expected it could quadruple its sales to $500 million by 1992.

By the middle of 1988, Gandalf had produced the market's first intelligent high-speed networking multiplexer or signal concentrator: StreamLine 45. Bailey began eyeing a possible takeover of a British company called Case Group PLC, with a view to becoming the largest independent manufacturer of its kind anywhere in the world. In May of that year, Gandalf made a $124.2 million offer. Bailey said Case would benefit from Gandalf's proven management abilities.

That year, Bailey was still predicting an ambitious $500 million in sales by 1992. He had been a financial manager with several companies in Britain before joining Gandalf as executive vice-president and CEO in 1982. He figured Gandalf could reach its target either by acquiring Case or by buying up several smaller technology companies. Gandalf was anxious to achieve a higher international profile; Bailey figured buying Case would do it.

Firms turn to acquisitions to help meet revenue targets.

However, it wasn't to be. After two months of tough negotiations, Gandalf lost the fight to Dowty Group PLC, another large British manufacturer of advanced electronic systems for aerospace, defence, mining and industrial uses that bought Case for $145 million.

In 1990, Gandalf lost $10.6 million. Cunningham returned as president after a two-year hiatus that had seen Gandalf lose money, and enforced a series of changes at the company to restore it to profitability: writedowns, layoffs and plant closings were among the tough measures. In retrospect, Cunningham said later, the company had focused on the wrong goals in his absence, embarking on too many ambitious development programs with other companies that, in the end, were too expensive. It was around this time that the company relocated from Manotick to its present location in an industrial park on Colonnade Road in Nepean, outside Ottawa. "Before that we were kind of scattered in several different buildings," says Burgess.

Another of Cunningham's moves, upon his return, was to acquire Infotron Systems Corp., an $85 million, New Jersey-based company that manufactured WANs. He was anticipating that Gandalf would need to change its tactics if it wanted to compete effectively — it would have to move into wide area networking. But with no previous background in WAN, there was no way Gandalf could enter the market fast enough unless it bought existing technology. Cunningham's solution: Buy Infotron.

Gandalf acquired 100 percent of the company, and Cunningham incorporated Infotron's WAN products into Gandalf's own technology base, enabling the company to produce a complex, fast modem, the Infotron 2000, which allows video, voice and data communication (ISDN). Cunningham was sure he could merge the two money-losing companies — Infotron lost $33.3 million in 1990 — and turn a profit. The idea was to attract larger and larger customers.

In 1991, Gandalf was awarded a US patent for its 2050 Network Communication Server QUIC Bus, designed to integrate cell relay and circuit switching on a common bus with no single point of failure. Following the merger with Infotron, Gandalf began to market the Infotron 2000 line of products, which integrate voice, LANs, imaging, video and data networking. A year later, they produced the industry's first integral, routed Ethernet-to-Token Ring internetworking solution, the Access Router. Merging with Infotron brought an array of new possibilities to Gandalf's existing technology base. At least six new products have been released since the merger, including the latest "super modems," reducing the innovation process by some 40 percent from conception to commercialization. (In 1993, Gandalf's new innovation, ASIC technology, created the world's smallest, fastest, cheapest local Ethernet workgroup microBridge.)

In September 1992, Brian Hedges — who had been Gandalf's chief financial officer for eleven years before leaving in 1988 for a four-year stint as chief financial officer at Teleglobe Inc. in Montreal — returned as president and CEO, lured back by Cunningham in an effort to make the company more profitable. (The company had been losing money for the previous two years, and the Infotron merger had been expensive. By the end of 1992, the company had lost $43 million in two years.) It was time for a change. Hedges laid off 100 employees just after his return to Gandalf.

When things get out of control, firms often bring in a "new broom" to clean up the place.

Burgess says Hedges had the right strengths for what the company needed then. On the edge of a turnaround, they needed someone with a strong operational sense and a finance background. "We really needed to inject change into the company," says Burgess, who describes Hedges as "a change man" who had both feet planted firmly on the ground. "He [wasn't] one of these airy-fairy, let's-all-hug-each-other-and-it'll-be-all-right guys," she says.

Although the Infotron merger was costly, and resulted in hundreds of lost jobs — most of them in the United States, some in Ottawa — in the long run, it gave the company a new edge, a strong R&D push and new technology. Gandalf could now combine its existing LAN technology with Infotron's WAN, enabling the company to offer customers a wide range of products allowing computers to communicate with each other around the globe. Gandalf also thinks the merger gave them a head start on ATM products, a technology that experts are predicting will be the new wave in the future.

The Infotron merger signaled Gandalf's new approach to the market: much broader than before. The reasoning was that this approach suits the future trend towards multimedia networks that will be so popular.

In 1993, Gandalf defined its target market as custom-made telecommunications products — modems, multiplexers, bridges, gateways, terminal adaptors — for companies whose employees often need to retrieve and send information to and from the company at large distances. That includes companies that need to communicate with branch offices, employees who work at home, and telecommuters. Hedges expected this focus to differentiate Gandalf from its competitors. The remote access market, worth about $1 billion now, is expected to more than double within the next five years.

In 1993, Gandalf made a final public offering of 12 million shares, taking in $48 million. And a year before that, Gandalf released a series of LANLine interconnectivity bridges in an attempt to gain complete dominance in the bridge market. So far, it is making significant inroads.

But despite these successes, the company continues to be plagued by losses. Although Hedges had initially been hired back by Cunningham in 1992 to restore the company to profitability, his two-year stint in the driver's seat was less than successful. In June of 1994 Gandalf announced a $47 million loss for the fiscal year; Hedges was replaced by Thomas Vassiliades, a recently retired veteran of the telecommunications industry. Changes in leadership are common after periods of heavy losses.

Gandalf is in transition now, says Burgess. "We're transitioning from an old product set to a new product set. So we have these legacy products that have a whole bunch of competitors; we have these networking products that are current, that have a whole set of competitors; and then we have this new place we're going to — we don't have products yet, but we know who our competitors are going to be. Most of them will be American companies."

The company's strategy, initially firmed up by Hedges upon his arrival, is to emphasize the remote access market, keeping a vigilant eye on trends in business, technology and tariffs. Its stated goal is to develop remote access solutions that help customers define and implement cost-effective networks, integrating voice, data, video, fax and LAN traffic — a multimedia approach that reduces network costs and simplifies network management.

"There were no competitors in the market initially," says Burgess. "The local data set and the data switch were both invented by Gandalf. We had the entire market to ourselves at first. Later on, the competitors came in — Develcon in particular with a local data set. Dataswitch companies sprung up. It took them a couple of years to catch on. Those days are over, though. It's way more competitive now."

> Creating a clear strategy as the basis for rationalization is essential to get everyone moving in the same direction.

Gandalf's products are now marketed and distributed in 90 different countries. More than half of its revenues are generated by the American market. The UK, Netherlands and France account for a further 36 percent of annual revenues, with the remaining revenue coming from smaller operations in other European countries, as well as Canada, Central and South Africa, the Middle East and the Pacific Rim. Customers are most often information technology users who view networks as a way to improve productivity and boost profits. Worldwide, they include financial services and health care organizations, government agencies and departments, universities, utility companies and similar service and industrial corporations. Gandalf maintains a 20 to 30 percent market share in local and medium distance modems; about 20 to 25 percent in dataswitches; and for multiplexers, about 2 percent. "We're number one in remote bridges," says Burgess. "The markets we're big in now won't grow, because they're decaying. The new market segment we're attacking is the remote access market — products designed to bring in traffic from remote offices or people who are working at home."

Spending on research and development now hovers around 11 percent of the company's annual revenues, or $20 million out of revenues of $160 million. Gandalf received no government subsidies or grants in the beginning, says Burgess — nor did they apply for any. Even today, she says, "I don't think we're using the government to the extent that we could be. It's a lot of work. You'd almost have to dedicate a person to finding and applying for these grants. There's a lot of bureaucracy, paperwork."

> To respond to rapid turnarounds needed by technology firms, governments will have to streamline their procedures to remain relevant.

Setbacks and Mistakes

Gandalf acquired Infotron to gain access to complementary existing technologies and respond more quickly to the market's increasing demand. In the process of the acquisition, Gandalf spent considerable sums simply moving its production from New Jersey to Nepean. Staff had to be retrained for the newer technologies; legal fees ate up a hefty chunk of cash; the net result was an unexpected decrease in revenues, and layoffs of some 300 staff.

Not having a clearly identified niche to sell into has probably hurt Gandalf's bottom line over the years. The newly defined focus on remote access products is the first truly narrow market they've focused on. Starting in 1985, the company launched about 15 different products in only seven or eight years. Between 1989 and 1993 it lost some $40 million.

Its challenge now is to erase any negative remnants of its past image from the minds of customers and investors, and re-establish itself as a determined, focused leader in the telecommunications field.

> With short product life cycles and pressure to be first to market, firms often get into alliances with other firms having know-how that is needed to accelerate product development. This approach is sometimes known as making a "buy" rather than "make" decision.

Risks and Choices

Something that will always be a risk for Gandalf is its large array of products: more than 20 new products have evolved from the original modem they designed in the early 1970s. The result is that the company has found it necessary to cancel some product lines, rendering them obsolete. Because of this, Gandalf's customers are often limited to those who are willing and able to pay for whole new generations of products. This rules out some potential business.

Gandalf's greatest challenge for the immediate future, says Burgess, will be developing alternate channels of distribution in North America. "Our direct sales force is really good when we're selling a systems solution," she says. "But technology has become more standardized, so basically any idiot can install it now. It lends itself to being sold through alternative distributors, such as distributors and value-added resellers, telephone companies, computer companies — other people selling and installing your product for you.

"A related issue is that the technology has a life cycle that's getting shorter and shorter," she says. "If you have a technological edge, it's for about six months at best. So the key now to success isn't necessarily the technology. You still have to have the technology, but that's the no-brainer part. The key is getting it out there, distributing it. You only have a six-month window to get the technology out there, you better get as much out there in that six-month window as you can. You need a sophisticated network of distributors and a strong sales force to do that. And we haven't developed the alternate channel distribution in North America to the extent that it should be. That's our biggest short-term challenge." As a result, she says, revenue levels are not where they should be in North America.

The Outlook for the Future

When Gandalf's marketing wizards look into the future, they see an increasing demand for new, innovative products that will provide better and better communication bridges; for example, they see demand for mobile systems in the wireless industry rising dramatically as less pricey devices make new applications more economical and practical. The ultimate goal is to turn Gandalf into the industry standard in video-conferencing, graphics transmission and similar functions. The market also looks good for Gandalf's ISDN product line. Gandalf's Netherland subsidiary predicts that by the year 2000, about 60 percent of all investment in information technology will be in software and services, up from some 30 percent in 1990.

Gandalf expects the networking business to expand because of three main trends: closer business relationships between suppliers and customers, requiring companies to extend their networks; more people working at home and communicating with their companies via modems; and deregulation of telephone companies, causing monopolies to dissolve and new competitors to enter the market, which in turn will require more diverse services and more businesses expanding their networks. In 1992, analysts estimated the telecommunications market to be around $6 billion, and it is still growing at a rate of about 36 percent annually.

Remote access will play a large role in that market, and Gandalf wants to have its foot through that door ahead of the pack. "We're focusing all of our efforts on making products that let the person get connected to a main computer centre in a way that is just as sophisticated as if he was directly connected, without any degradation from the distance of the lines he's had to work over," says Burgess. "And we're doing it so that you can use the cheapest services that are provided by the telephone company. This is our chosen target market. The product, called Expressway, is designed, we've announced it, and we'll start to ship it February [1994]. It consists of a centralized concentrator that you would put into a computer centre, and then there are all these remote devices that are really low cost and small that go into the remote offices."

That segment of the market is worth about $300 million now and projected to be $1 billion in 1997, says Burgess. And, she adds, it's one that has no dominant players yet. "We'll be one of the first in the market. And it plays nicely to our strength as a company, because of the merger with Infotron — the remote access thing requires the ability to do both (wide area networking and local area networking) because most of those remote offices have LANs, and it's quite hard to take LAN traffic and cram it down WAN pipes," she says. "LAN traffic goes at megabits per second, and WAN goes at kilobytes per second. So you have to do sophisticated things with the traffic, like compression, to squeeze it down that pipe in a way that doesn't degrade what the customer is seeing.

"Because of the merger, we now have the skills in-house to address that market, and a wide range of technology to support it," she adds. "The breadth of technology understanding that we have over 20 years should give us a good edge."

The Competition

Gandalf's products usually require one to three years to move from the conception stage to commercialization. The strategy is to introduce each new product to the market as swiftly as possible, before competitors can, so as to enable Gandalf to maintain and even expand its market niche with LAN and WAN products. The company plans to deal with competition by incorporating new technologies into Gandalf's existing products to make a new generation of faster, more complex and reliable modems — the Infotron 2000, for example, combines video, voice and data networking in one product, and relies on highly sophisticated Intel microprocessors.

When it first started up, Gandalf had no competitors. By the early 1970s, Gandalf's only competitors were other data switch makers. By the mid-80s, both large and small communications companies of all sorts were beginning to move in on Gandalf's market in data services. The importance of transporting data between computers had gradually dawned on everyone in the industry as a key feature of information technology for the future. Some of Gandalf's key competitors were Develcon Electronics Ltd. of Saskatchewan, and local up-and-comers such as Mitel Corp.

By the late 1980s, Gandalf's chief competitor in the United States was Timeplex Inc. In 1988, that company was taken over by computer giant Unisys Corp., prompting executives at Gandalf to become even more wary of the trend in their industry towards larger firms swallowing smaller ones. Avoiding this fate was part of Gandalf's motivation to pursue Case that same year.

Because of intense competition that began to crowd Gandalf's market in the last five years, the company has been forced to cut its prices on some products by as much as 40 percent.

Management Strategy

In order to keep a tight focus on its products, market niches and goals, Gandalf's strategy is to divide its operations into three distinct areas. One is a branch responsible for R&D, manufacturing and marketing. Based in Nepean, this division focuses on communication products and systems. A second branch is network systems sales and service, which provides solutions for local and wide area networking, mobile data networks and ISDN access. The company sells its mobile data systems to taxi and similar transportation markets. The third division is network component products; its purpose is to ensure that products — such as modems, terminal adaptors, bridges and transmitters that are bought by other companies for resale — reach the market place as quickly as possible, with highly competitive prices.

In order to stay competitive and be the first to react to market demand, Gandalf managers continually make the rounds of conferences, trade shows and universities, always keeping an eye out for the latest innovation in telecommunications. To that end, they also hire market researchers to monitor their competitors' latest designs.

The ability first to churn out constant innovations to existing technologies and then to manage the accompanying growth successfully has been, and will continue to be, a large part of the reason for Gandalf's success. "I remember when I first worked here, we used to come out with a new product each month," says Burgess. "Organizationally, it certainly is crazed. You're growing so quickly and you don't have a structure to make anything efficient, so you end up reinventing the wheel over and over again in process terms. So there are lots of inefficiencies. I think, considering, we managed it quite well. But it has left us with a legacy that's a bit of a problem in terms of fixing processes."

That's where strategies to effectively manage creative people who don't want to be told what to do become important. "This company has virtually no formal processes," says Burgess. "Everything is done informally because that's how we survived the high growth phases — we *just did it*. So we have 1,400 people *just doing it*. And they all have really good intentions, and they're all really keen, but they're not all necessarily working in sync."

> Firms use international quality standards to impose discipline on their skilled workers. Such standards are becoming a basis for ensuring international competitiveness.

Recently, she says, international quality standards helped Gandalf to fix that problem in particular. "We're moving more and more toward measuring up to certain preset industry standards. But we still have this legacy of people wanting to do it informally. It's hard to get them to accept processes.

"Also, we have a highly educated staff. Most of the staff here are professional. They don't like rules. They're creative. They want to push the envelope as much as they can."

Gandalf's greatest strength, says Burgess, is its history — 20 years in the market have taught its executives hard lessons and garnered the company a solid base of loyal customers. "We also have an international distribution infrastructure for pre- and post-sales support, as well as direct sales, which is quite an expensive thing to set up," Burgess points out. "I think if you look at the competition we have a better infrastructure than they do for distribution."

But a good distribution system still needs new products, as the $47 million loss announced by Gandalf in June 1994 shows. Burgess said sales of the company's older products fell off more quickly than expected, especially in North America, though sales of newer products were starting to build up. As noted earlier, in May 1994, Brian Hedges was succeeded by Thomas Vassiliades as president; Vassiliades had recently retired as CEO of Bell Atlantic Systems Inc. Given the circumstances, Burgess admitted that further restructuring could not be ruled out. "The only norm now is change," she said to the *Globe and Mail* on June 4.

16

International Submarine Engineering: From the Depths of the Oceans to Outer Space

ISE is a firm whose success is linked directly to the sustained vision of its founder that the core technology has to be used to create related products that can migrate to a number of well-defined markets.

International Submarine Engineering, founded in 1974 and now a $10 million company, brings to life the kind of technology that most sci-fi fans only dream exists. Originally a manufacturer of unmanned submarines, ISE's first exposure to outer space technology came in 1993, when it designed a robotic arm for the Canadian Space Agency (CSA). The arm and its controlling software, developed for $1 million, are the result of a four-year contract with the CSA. This contract requires ISE to develop a ground-based control system of space-based robots for testing. A group of 10 ISE staffers spent most of 1992 creating the robotic arm and controlling software. This area of research has now become the company's biggest focus.

In June of 1993, ISE announced it was teaming up with Spar Aerospace — the leading Canadian manufacturer of space-based products that built the US space shuttle remote manipulator system, better known as Canadarm — to pursue opportunities in space robotic projects. The shared goal is to get sophisticated robots into space, eventually replacing manned vehicles.

The reason this is big news is that the 20-year-old company, prior to this contract, had been involved almost exclusively in designing and manufacturing unmanned submarines; space technology marks a pivotal departure not so much in technology, but in market. Jim McFarlane, the founder and president of ISE, has a master's degree in naval architecture and four honorary doctorates — two in engineering, one in science and one in military science — and spent 18 years in the navy before he started up the company. He left the navy in 1971 with the idea of building mansubs and started ISE in 1974.

By that time, McFarlane had decided to build remotely operated undersea vehicles (ROVs) for the offshore petroleum industry. "The 1960s were a very vigorous period in technological development," he says. "The world switched from sea and rail to air travel; mansubs were developed at the end of the 1960s; there was one tethered vehicle developed during that time by the US navy; and then in 1974, when the first centipedes were coming out, there was a proliferation of people trying to make tethered vehicles. It was possible and practical because of the high-density electronics that had become available."

Since then, ISE has manufactured more than 200 remotely operated vehicles, and still markets them worldwide. In the early 1980s, the development of microprocessors allowed ISE to launch its development of autonomous underwater vehicles (AUVs), the world's first. In 1982, ISE Research Ltd., a wholly owned subsidiary of ISE, was created to continue the development of AUVs, manned submersibles, and other subsea systems for various uses.

The experience ISE accumulated in developing remote systems that could function in the rugged environment of inner space eventually allowed ISE to apply that technology to space-based projects. The company is currently involved in a contract with the CSA to develop autonomous robot technology applicable to the Space Station Freedom project.

The company boasts a history dotted with high-profile expeditions into dangerous situations. ISE technology helped cap an oil well blowout in Mexico. It recovered the black box of the Air India jet that crashed off the coast of Ireland and killed 329 people. In 1982, it defuelled one of the reactors of the Three Mile Island disaster, using a robotic arm to dismantle and retrieve contaminated pieces of the damaged core of the power plant so that people could go in and finish the clean-up job. And an ISE subsidiary took part in the search for the Titanic.

McFarlane attributes most of his company's success to his own marketing acumen and sense of purpose. "You have to have a mission in order to do anything," he says. "Otherwise it's just a laboratory curiosity.

"I say I'm going to build an autonomous vehicle. So I build it. Suppose I say it's going to be six or seven feet long and two in diameter. So now I've built this thing and I go running up and down the street saying, 'Look at my autonomous vehicle.' People will say, 'Who the hell cares? What's the mission? What are you going to use it for?'

"Whether you intend to use it for surveying," he says, "or for looking at polymetallic sulfide deposits, or even if you're going to use it for science — the question is, what kind of science? You still have to know what the mission is, and characterize the subsystems of your vehicle to respond to that mission. Otherwise it's just a curiosity."

That's a mistake that happens all too often in engineering, says McFarlane. "Some person gets a great idea. They have no clear notion about what the hell the market is. And then they go and build the thing anyway, and have no idea why nobody wants it."

> A sustained sense of purpose, that is a vision or mission, is fundamental to success because it focuses the efforts of the firm.

History and Milestones

From the moment of his company's inception — which was funded, McFarlane says, with his "personal nickels and dimes" — McFarlane concentrated on constructing tethered underwater vehicles. These were connected to the water surface by an electronic umbilical cord, and used mostly for inspecting underwater pipelines and to support drilling operations in the offshore petroleum industry. The cord facilitated the operation of the submersible and its manipulator arm from a closed-circuit TV.

"Developments in the ocean in the late 1960s were primarily driven by the offshore petroleum industry, both in the Gulf of Mexico and the North Sea," says McFarlane. ISE's first job was pipeline inspection using tethered, remotely operated vehicles. "I knew the primary market was the Gulf of Mexico because that's where head offices were for a bunch of oil companies. But the water wasn't as deep there, and the area was less technology-intensive than the North Sea. So I went to the North Sea first, then to the Gulf of Mexico," McFarlane says. "I knew I had market position." ISE's first major contract was in the North Sea, towards the end of its first year. By that time, the company employed five people.

Throughout the course of its history, ISE — which now employs 70 people — has produced a wide array of undersea products, sold almost 200 different systems, and developed an international reputation for design and manufacture of underwater marine equipment.

ISE's presence is part of the reason that the area of Vancouver in which it resides — Port Moody — is sometimes also known by those in the industry as "blue-water Silicon Valley."

ISE produces remotely operated undersea vehicles, manned submarines, autonomous remote controlled submarines, tourist submarines, imaging sonar systems, and has now made the leap from deep sea robotics to space robotics. Its first tethered vehicle was produced in 1975; its major customers have traditionally been both the military and oil companies, which need submersibles for exploration, laying and repairing pipelines, and tool recovery. ISE's primary focus for the first decade or so was remotely operated undersea vehicles, but McFarlane says that began to shift a little about 11 years ago with the introduction of microprocessors.

> A product migration strategy ensures that the technology can be adapted to a number of related applications.

"It became possible to control autonomous vehicles — semi-submersibles and fully submersible autonomous vehicles." Autonomous vehicles operate without the often cumbersome umbilical cord featured on the remotely operated ones. ISE was first into that market, and remains the only company in the world that offers autonomous vehicles. "We started early," says McFarlane; no one could catch them. He's still waiting for the competition to try to edge its way in.

"I fully expect the thundering herd syndrome to set in again in that market," he says. "If you look at the ROV market, for example, by 1974 or 1975 it was saturated. By 1978, mansubs were blown right off the face of the earth. There's only a few left in the world, most of them in government use. Next there were tethered vehicles in pipeline inspection, which used to be a job for mansubs. We blew them out of the water and saturated the pipeline inspection market.

"So then we went to the drilling support market. Of course, now with oil at $18 a barrel it isn't a big business any more. With this kind of technology, the market goes up and down and starts to curve over ... The bottom line is, you have to put something in right behind it and get going."

One of ISE's subsequent major contracts involved designing and producing a series of Dolphins (Deep Ocean Logging Profiler Hydrographic Instrumentation and Navigation): unmanned, 3,200-kilogram, diesel-engined semi-submersibles that can travel up to 16 knots underwater and are controlled by radio signals. Commissioned by the Canadian military to detect seabed mines, the Dolphin was developed for use in hydrographic and bathymetric surveys in rough waters (to map the sea floor). ISE has sold five of these to the Canadian Hydrographic Service for underwater mapping, and two to the US Navy.

In 1984, McFarlane launched production of AUVs — a second generation of torpedo-shaped, underwater vehicles that, unlike their ancestors, operated without umbilical cords. These updated machines worked independently, allowing for increased flexibility.

Another of ISE's early products was a remotely operated craft controlled through a fibre optics tether, completed in 1986. This 50-hp, US$1.7 million unit, christened Hydra-AT, was built for Shell Oil's deep drilling operations in the Gulf of Mexico. It could dive to an unprecedented depth of 2,000 metres, attached to its mother ship by a thin umbilical cord. Controllers sat in the ship, sending the ROV commands via light signals along the tether, which contained fibre optics filaments. By 1986, ISE had built some 150 unmanned underwater craft, and its customer list included the Canadian Navy, the Brazilian government, and some of the largest diving outfits in the US and Canada.

Later in 1986, an ISE robot arm was activated by an operator in China in an event later called the world's longest handshake. Around the same time, ISE began designing a 34-passenger submersible for International Union Resources Ltd., which wanted to build a submarine for tourism.

Then, towards the end of the year, ISE became a principal Canadian shareholder in a $50 million nuclear submarine project that was due to enter service off Canada's east coast in 1989. The hull was being built in France and the power plant supplied by an Ottawa company, Energy Conservation Systems Inc. (ECS). This venture marked a departure for ISE, since the submarine was designed to carry up to 12 people, including divers, who would be able to use it as an "underwater hotel" during lengthy projects in depths as great as 300 metres. The vessel was being built for International Submarine Transportation Systems (STS) of Halifax, Nova Scotia; ECS of Ottawa and ISE were to hold, between them, 50 percent of STS's shares, with the balance to be split between Comex — a large French diving company — and a French government agency involved in deep-sea research.

In 1987, ISE became a partner in a consortium led by Canadian Shipbuilding & Engineering Ltd. of Toronto to build 12 maritime patrol vessels for the government at a cost of $540 million. The vessels, commissioned by the federal government, were to be used to hunt underwater mines by the naval reserve of the Canadian Forces.

The year 1988 heralded the arrival of Hysub 5000 (whose funding was co-sponsored by the Department of National Defence, which expected it might need them in the future for mine countermeasures), the world's first unmanned submersible that could dive to a depth of 5,000 metres, more than twice as far as any previous similar vessel had ever plunged. Around that time, industry observers started to hint that Canada's lead in developing and marketing remotely operated vehicles was in jeopardy due to a lack of financing and coordination. Up to this point, it was estimated that Canada had made one-third of all non-military ROVs in the world, including the world's first large radio-controlled robots. Those in the industry began to complain that this success had gone largely unnoticed in the global community, jeopardizing Canada's chances for marketing the technology. The long-term future, however, still looked bright — oil prices were predicted to rise, bringing the traditional ROV market back to life by supporting offshore drilling. Oceanographers were beginning to take note of the ROVs' potential in research; the robots' potential for finding and disabling mines, it was expected, would eventually create a market worth a total of $5 billion.

That same year, McFarlane acknowledged that the advent of the autonomous vehicle meant ROVs had been reduced to a restricted market, with diminished recognition of their value. But he also predicted that within three or four years they would still be profitable, while the technology developed from them would be useful for other applications.

In 1989, ISE was still the only company in the world manufacturing autonomous underwater vehicles. McFarlane speculated that the market would probably lure multinationals soon. That year ISE was working on a contract with AT&T to design a deep-diving vehicle to repair and inspect transoceanic telephone cables, and it had just begun to branch into the space market with NASA, studying the use of robotics in space. The firm's first involvement with space technology also occurred in 1989.

In 1991, the federal Department of Industry, Science and Technology awarded ISE an $800,000 contract to develop the autonomous Robotic Refuelling system over a three-and-a-half-year period. The Wright-Patterson Air Force Base in Dayton, Ohio, contributed matching funds. Using computer vision, the system was designed to identify a specific aircraft and detect its position. A manipulator would guide a computer vision camera to inspect specific sections of the aircraft and then connect a refuelling nozzle to it. This system is much more precise and effective than current approaches. Funding for this project was provided by the Defence Industry Productivity Program.

In 1992, ISE branched out into the UK, opening up ISE (UK) Ltd., a sales support services subsidiary. The plan was to help boost sales to oil companies active in the North Sea, which McFarlane perceived was a high activity area where ISE's remotely operated vehicles were used to support drilling operations and sea floor cleanup. That year ISE had 90 employees and expected $10 million in sales.

> Even firms with well-established international reputations need, at some point, a presence in the markets they wish to penetrate.

By the summer of 1993, ISE had teamed up with Spar Aerospace Ltd. to continue to explore business opportunities in space robotics. The aim of the partnership was to put sophisticated robots into space, promoting the use of robotics over manned vehicles there. The alliance was the next step up from ISE's subcontract with Spar, reached months earlier, to develop a ground-based remote control system for space robotics, and on its research contracts with the CSA to build mock-up ground-base control systems. The agreement was mutually beneficial. Said Eric Jackson, ISE manager of space programs, at the time: "I don't think anybody can argue that Spar isn't the foremost company in the world when it comes to space robots ... But Spar has only built 10 robots. ISE has built 250. We sure know how to build robots." About 12 ISE employees began working full-time on the project with members of Spar's Advanced Technology Systems Group.

> A strategic alliance can be a win-win situation in that it brings the complementary skills of each partner to bear on the development of a specific product.

Later that summer, ISE completed construction of a robotic arm for the CSA. The arm is a replica of one of the 80-inch arms — similar to, but smaller than, the Canadarm, used on the US space shuttle — to be used for testing purposes for another year or two before it is sent to the space agency.

Contracts are coming more often now from the government and military, says McFarlane — a shift from the company's previous emphasis on oil and gas companies. It's all the same to McFarlane, who is happy to keep producing products based on market requirements. "We create the vehicle to suit the mission," he says. "Over the years, vehicles have come in all shapes and sizes." ISE products have been variously described as resembling underwater sleds, strange sub-sea spacecraft and even ocean-friendly versions of

R2D2. "The key is imaginatively adapting basic technology to a spectrum of needs."

What remains to be seen is whether or not the company will now shift its focus almost exclusively to space technology, since that seems to be the direction in which it's currently headed. McFarlane is quick to point out the similarity of the technology; robotics is robotics, whether in space or under water. However, the space technology market looks lucrative. The Canadian government has pledged a total commitment of $1.3 billion to it for this decade, $750 million of which has already been spent on designing robotics for the space station. In the United States, NASA has spent nine years and $8.75 billion on the $31 billion station, under budgetary review by Clinton in late 1993.

Despite the transition from submersibles to space robotics, ISE has always retained a tight focus on its market; regardless of their strange appearances and diverse uses, its products have always been linked by a common thread.

"First of all," says McFarlane, "if you look at the line between autonomous and tethered vehicles, you see they have commonality in parts and control.

> Many technology-intensive firms are undercapitalized and have to fund the development of their products through a variety of approaches ranging from government support to getting the customer to pay.

"If you look at the space projects we're doing — power management, autonomous robotics, trajectory planning and control — and then you look at the underwater vehicle, it's the same stuff. The area we aren't knowledgeable in, with space technology, is the materials. But with the whole business of teleoperators and control, the most experience in the world is in the subsea industries. In those industries, people have been going to work under water every day for 20 years, whereas the space guys have seldom gone to work at all."

ISE's research and development strategy is to try to make the customer fund as much of it as possible. The major drawback is the time it takes to get from development to production, says McFarlane. "This is a hard one to call. For any major vehicle being integrated for the first time, we have to sell prototypes. We can't afford to build prototypes and then have them sit around while we go and build something else. We have to essentially bet the store on these developments. We're trying to get our customers to fund the R&D, so spending varies."

One problem with this method, aside from the fact that R&D spending can fluctuate wildly, is what it does to the bottom line. "All the money you spend gets expensed and makes your balance sheet look like a dog," says McFarlane, who thinks the biggest problem facing small high-tech businesses today is underfunding. "What you're doing is bootstrapping yourself, project to project, because you're undercapitalized and you can't go to a bank and get the money. That's another reason you have to have a mission-driven product and go and sell it: You don't have the deep pockets to go and develop something that doesn't have an application."

Setbacks and Mistakes

McFarlane also says any major setbacks his company has suffered are directly attributable to lack of capital, resulting in the inability to follow through on unanticipated opportunities.

"Any Canadian company up to 150 or 200 employees is undercapitalized, including us," he says. "They just don't have the working capital. You have to keep moving when you see the chance, one way or another, and you can't go to banks; they're not risk lenders. Pork and beans, fine. But if you're talking about tethered remotely operated vehicles, forget it."

The other problem, says McFarlane, is the way balance sheets are set up and how banks subsequently perceive them. "Drawings resulting from R&D are expensive in the year they're done, in technology companies," he says. "What that means is all the money spent on research and development looks like it's wasted, because you have nothing concrete to show for it. Drawings that you're subsequently going to work from are useless from an accounting perspective."

So, he concludes, all small high-tech companies contend with the millstone of not being able to ascribe any value to the research and development they've spent a year or more doing. "That means the assets are grossly understated," says McFarlane, "but from a banking point of view it looks like they've got nothing."

What's the answer? "If we're ever going to grow major high-tech companies here in Canada, we're going to have to change our attitude a little bit," he says. "I'm not talking about grants. I'm talking about some way to adjust the accounting rules — ways to make sure there is working capital available. There are companies that can sell stuff, but then find they can't go out and build it, because they don't have the money to finance it. We're losing jobs in Canada because of this problem."

This is a problem McFarlane says has hurt ISE several times. "If you really strip the guts of it out," he says, "lack of sufficient working capital is what limits your ability to respond to the market place. I've seen opportunities I couldn't respond to, often, for that reason. It means you can't have things in inventory — major pieces of equipment — whereas company XYZ in England has government support and can build a piece of equipment and hold it on the shelf. Then a job for drilling support equipment comes up and they need it in 4 weeks, the bill period is 26 weeks, and you're dead in the water. There's no way you can respond to that sort of thing. It's an industrial strategy in the national sense — we don't have one."

The United States, says McFarlane, is a good example. "It's industrially more developed than Canada. It's not quite such a branch plant mentality. Per capita, 10 times as many companies in the US reach $100 million as compared to Canada. That's 100 times as many $100 million companies. That's got to tell you something."

Risks and Choices

After you've created a market, says McFarlane, you have to concentrate on differentiating your product in terms of either capability or price, which is one of the reasons he thinks ISE's ROVs and AUVs have succeeded in staking such a solid claim within their respective markets.

"If you can't do that, you're dead," he says. "The product won't strike anyone as being unique. Part of that differentiation occurs because of position, too. Market position means you are at the point of initiation, essentially. Your name is clearly understood like Kleenex and Xerox, not no-name." Similarly, in the world of submarines and now, increasingly, space technology, the name ISE has become synonymous with leading-edge robotics technology.

"It can be risky," says McFarlane. "But there's risk in status quo, too. Ask IBM. The time rate of change today is a lot faster than it was at one time. So you can't rest on your laurels. You have to be flexible, and you have to have flat structures in terms of management. You have to work on matrix type management — project management."

The Outlook for the Future

Looking to the year 2000, McFarlane sees space and sub-sea technology growing even more autonomous than it is now.

"As things evolve," he says, "it's like Darwinism. You start with something, it moves out, genetically it changes a little. The DNA is close, but there are some differences there. The other thing is, we want to get as much leverage as possible off the technology we have in order to shorten the time it takes to develop something. The more developed capability we have, the more benefit we'll get off the leverage."

Turbulent times of rapid growth are over now for some ISE products, but just beginning for others. "In the ROV business, we've flattened out. It's a fairly mature industry. In the autonomous vehicle business, we're just on the upslope from the bottom," says McFarlane.

Autonomous vehicles will grow in two directions: science and military. "The attributes will be mission-driven," says McFarlane. "The vehicle is the platform, and the things it's going to deliver will be mounted upon it."

Smaller, more portable autonomous vehicles with increased capabilities are the market of the future; McFarlane will make sure ISE is first into that market. "That's the aim of the exercise," he says.

The primary competing ground will be the United States, says McFarlane, because of their substantially larger industrial base. "As soon as it's perceived that there's an interested market of substance, people will begin to compete."

The Competition

For ISE, the key to success when competing with much larger firms has been always having a solid head start.

"In any technology you have to have market position or you can't move," says McFarlane. "There are lots of demonstrations of that. Look at Apple. Look at Microsoft. They had no management experience, bad haircuts, no dress pants, nothing. But they had market position.

"Later — four or five years later — when AT&T tried to move into that market, they couldn't. They had all the management, all the marketing experience, all the everything. But they didn't have market position." ISE has narrowly avoided being crowded out by manufacturing giants in much the same way. "Initially, when I started to develop these vehicles I had market position," says McFarlane. "But the thundering herd syndrome set in fairly shortly — within six months of having started, more than 30 other companies in the world were building a tethered vehicle."

Still, he says, ISE managed to grab a substantial section of the market. "We had as much as half at one point," says McFarlane. "And as the market began to mature, there was a concentration of power. The market can't support everybody. Some — lots — of our competitors faded out." Today, ISE is one of only three companies in the world in the tethered vehicle business.

Management Strategy

McFarlane can't emphasize market position enough when it comes to discussing how to succeed in high-tech.

"When you look through a tube and see light at the end of the tunnel, you have no reference," he says, drawing an analogy. "You can see there's light but you can't tell how long the tube is. So, I can see that there are autonomous vehicles required, and I can see the applications of them a long way ahead, but I can't convince people that's so. Nevertheless, I have to have position when that happens, or I'm not going to play in the game."

He emphasizes the fine line between creating a market and grabbing a share in a pre-existing one. "You could argue a case, for instance in ROVs, that the issue was just market share," says McFarlane. "But we had to create the market for the ROVs first, and then they competed for share in those areas where they were working.

"Also, there were a whole bunch of new areas that were not practical with man submersibles that were taken on once ROVs were available — all sorts of things that couldn't be done before. That was market creation as well. And if you look at the autonomous vehicle, that was market creation too.

"There are analogies all over the place. If you look at PCs, when they first came out, the usual response was, 'Who needs it — I've got a calculator, my trusty slide rule, a typewriter.' But as time went on, the efficacy of the PC became obvious. Then the thundering herd syndrome set in."

The best way to manage a rapid growth phase, says McFarlane, is to use the principles of project management and avoid spending too much capital you can't afford on hiring dozens of new employees you can't support in the long run. "The more rapid the growth, the more I use subcontractors," he says. "The surest way to destroy yourself would be to try and accrue all this capability into your house. Your in-house capability has to lag the market. You have to have some perception of how big the market is going to be, therefore, so you don't go out and get a big number of people that in the long term you can't support." McFarlane says ISE has never seen a rapid growth phase that he couldn't successfully manage this way.

One way of dealing with rapid growth when capital is scarce is through alliances, including the use of subcontractors.

For companies struggling to get a growth phase to happen, McFarlane's words of wisdom emphasize the importance of understanding your market. "Make sure the product is market-driven," he says. "Because if it isn't, you're going to have a long time getting into business. You need market position. If somebody is already well into the market, unless you have a very easy way to differentiate your product, there's no way you can succeed." He also emphasizes keeping a tight focus on your goal. "You have to initially limit your objectives and have a strong focus on that objective," he says. "It's almost like having a mission."

Finally, taking risks when necessary is crucial to success.

"You have to be prepared to move," says McFarlane. "Accept unreasonable deadlines to exercise your craft. It is essential to exercise your craft, or you can't evolve. You can't pick up people and lay them down like you can in low-tech. You have to stay put until there is no alternative but to downsize. You really have to play brinkmanship in the most aggressive way.

"You have to have a poorly developed sense of fear."

17

Coping with the S Curve

The development and growth of firms generally follow a pattern approaching an S curve, which has four stages: the start-up, the introduction of the first products, rapid growth and then maturity leading to either renewal or demise.

In this chapter, patterns, lessons and principles tied to the growth of technology-intensive firms are explored through the case studies.

The Start-Up

The beginnings of a technology firm can vary widely. The original idea can come from anywhere — a summer job, for example, as was the case with Techware, or a PhD thesis, in the case of Mitel. In many instances, however, a firm begins with technology transferred from organized research, as was the case for Xillix.

A popular approach is to start off as a consulting firm because many entrepreneurs do not have a clear product in mind and the overheads are low, which reduces the risk. This is the way DY 4 and Dynapro got started. A related approach is the R&D company that undertakes contract research until a product idea surfaces. Both Vortek and Creo started this way. Cognos represents a similar situation. The firm originated in 1969 as Quasar Systems Limited, a computer consulting services company. Under the leadership of Michael Potter the firm moved on to develop an ongoing series of successful software packages that spurred its growth.

A variant on consulting is to begin as a systems integrator of other firms' products and components. Corel started in 1986 as a desktop publishing systems integrator. This meant that it was originally involved with both hardware and software. Realizing that the profits in hardware were limited since clones can be produced easily and cheaply, Corel opted for software development because of the proprietary advantages. Hence the development of CorelDraw, which launched the company.

The classic knowledge-based firm start-up situation is one in which the founders have developed a technology, but do not have a market for it. This is called the "technology-push" syndrome. A classic example is Vortek's powerful arc lamp. For the last 20 years the firm has been attempting to find an appropriate market niche for its technologically superior product.

The reverse situation is "market pull," that is, a situation where an entrepreneur acts on a clearly identified niche opportunity. Such was the origin of Instantel. A mining engineer expressed his frustration at the poor level of blast monitoring technology to a high-technology entrepreneur. The accepted approach to blast monitoring before Instantel entered the market was to record explosions on magnetic tape and then send the tape for analysis to a laboratory in the US. The 20- to 30-day delay in getting the results could be very costly if disputes arose over possible damage resulting from blasting levels. There was a need for a blast monitoring instrument that indicated any problems immediately. Doyle saw the opportunity, wrote a business plan, organized the necessary financing and launched the company.

Gandalf Technologies Inc. was also launched when a niche opportunity presented itself. Ottawa entrepreneurs Des Cunningham and Colin Patterson responded to the demanding requirements for an advanced data communications device set out by an engineer at the federal government's Communications Research Centre. This led to a low-cost, leading-edge data communications modem that was an attractive alternative to more complex and expensive devices.

And then there are the "product champions" who are so convinced of the merits of their ideas in a specific market that they are unstoppable. The most striking example of such an individual in recent years is Terry Matthews, who founded Newbridge Networks Corp. in 1986. Matthews saw the need for improved communications networks that would handle both voice and data at enhanced speeds. So convinced of this requirement was he that he invested some $14 million of his own money, money that he had made as co-founder with Michael Cowpland of a previously successful firm, Mitel Corp.

These examples show that there is no fixed formula for starting a knowledge-based firm. The entrepreneur usually begins with a promising technology and hopes to find a market. Sometimes he or she will start with a market opportunity and develop products that will sell in that market. There is also the stop-start approach of edging into a firm by ferreting out the right opportunity through consulting or research.

Similarly, there is no fixed pattern for the initial financing of a venture. All possible private and government sources of financing are used as need be. Usually, personal sources such as family and friends, mortgages and government grants are used in the earliest phases of product development. Once the concept is firm, possibly with a prototype, then more significant and formal sources, such as venture capital, large investors and the stock market, can come into play. In fields that are capital-intensive or where strict rules have to be followed, such as pharmaceuticals, all sources need to be tapped early because of the high costs and the length of time involved in bringing a product to market. Quadra Logic, for example, found that it needed major financing to bring its drug to market, and got it through large infusions of money from American Cyanamid and by putting its shares on the stock market.

Since financing is always an issue, there can be advantages in launching first products that are not capital intensive. There can be advantages as well in some technical fields, such as software, that do not require as much capital as other areas. It is also helpful when the entrepreneur can provide much of the start-up financing, as was the case with Terry Matthews at Newbridge and Michael Cowpland at Corel. However, as a general rule, entrepreneurs usually underestimate the time, and hence the resources, needed to bring the first product to market.

The public sector is at times a good market entry point for new advanced products that the private sector views as risky or unnecessary; governments and universities can have requirements that are ahead of those of industry. This is how Gandalf and Techware got started, for example.

The First Products

The first product launched is extremely important because it is the firm's first exposure to the market place. A well-received product can start the firm on its climb up the S curve, while a poorly received one can kill a new firm.

Consider the case of Newbridge. Its first multiplexer product was well received in 1987, and the firm began to grow rapidly. However, ensuing technical difficulties that were costly to repair, overextension of manufacturing capabilities and unfulfilled promises to clients severely curtailed the firm's growth. Newbridge has since recovered, and its stock, which had fallen to a low of about $4/share from about $20/share during its period of difficulties, was in the $70/share range by late 1993 after splitting two for one. Newbridge had the resources to recover, but such is not the case for many firms.

Corel began to grow rapidly when it launched its very successful CorelDraw in 1989. Subsequently, two improved versions of CorelDraw were introduced, one in 1990, the other in 1992, indicating a straightforward product migration strategy. In 1992, the firm also launched a new product, CorelSCSI, a software package that facilitates the integration of high performance peripherals required for graphics.

Cognos followed a similar path, but took much longer than Corel to make the transition to a software developer. This computer consulting firm developed advanced software packages which it began to introduce into the market some 10 years after it was founded in 1969. By 1981, Cognos had introduced three software packages. In 1982, it began to integrate these three products into a single standardized product called PowerHouse, which was very successful and launched the firm's growth phase.

DY 4 initially dabbled in consulting for three years before finding a product niche, the VME bus for military computer systems. The firm established itself through a first big contract with the Danish navy, followed by other major contracts with Raytheon and Rockwell. That first big contract with an established buyer is extremely important because it means credibility. To link unknowns (at the time) with established companies and organizations, like Dynapro with Molson Breweries, Techware with Stanford University, Creo with the Canada Centre for Remote Sensing and Gandalf with the Communications Research Centre, is beneficial in establishing an early market presence through the first sales.

Instantel had a clear market focus from the start. It made its mark in the well-established quarry blasting market by introducing a portable seismograph that offered real-time recording in the field of the energy released by a dynamite blast. This was a breakthrough product that caught the industry by surprise. The functionality of the product, the convenience of a real-time readout of a blast on site, was far superior to the delays incurred through the conventional laboratory analysis of tape recordings of blasts. Subsequent products also provided superior functionality.

Good products are very important. Corel found a market niche after a few tries, but Vortek is still looking for an appropriate market niche after 20 years. Having a clear market focus, as the examples of Newbridge and Instantel demonstrate, is key to successful product introduction. The introduction of the first product needs to be sustained by the introduction of related products. A well-thought-out product migration strategy is essential for growth.

Climbing the S Curve

As a firm starts to grow, managing growth becomes the central focus. The entrepreneurial approach has to give way to professional management. Firms that can't make this transition either fail or stagnate.

Newbridge illustrates this problem very well. As noted earlier, Terry Matthews, the founder, had a clear vision of the market and introduced products to serve that market. However, Matthews's marketing effort far outpaced product performance and the company got into trouble in 1990. The board had to bring in a chief operating officer — Peter Sommerer, a former Mitel employee — to take over some of Matthews's management functions. The firm, which had lost $12 million in 1991, returned to profitability in 1992 and continues to grow rapidly.

A similar situation occurred at DY 4 when its revenues plunged in 1989 due a misadventure in an unknown market — Indonesia. The two major investors, CBC Pension and Noranda, bailed the company out on the condition that the CEO, Garry Dool, step down. Dool returned in March 1993 to oversee the next phase of growth based on the funds made available by going public.

At Cognos, Michael Potter, the driving force behind the firm's early growth, decided to step aside as president in 1986 when the firm had reached $45 million in sales. The new president and chief operating officer, Thomas Csathy, was a seasoned veteran from IBM Canada and Burroughs Canada. Csathy's presidency lasted three years, during which time Cognos grew to the $100 million level, but not without difficulties that led to further management changes aimed at establishing a proper balance between maintaining an entrepreneurial spirit and professional management.

In anticipation of the need for professional management some firms get these managers before a crisis precipitates the requirement for a change at the top. For example, the founders of Xillix knew that they needed professional management early on and recruited

David Sutcliffe, an experienced high-tech manager, as their CEO. On the other hand, Karl Brakhaus was able to steer the steady growth of Dynapro from the beginning. He was helped in large part through the firm's alliance with Allen-Bradley. Now the firm is moving to its next phase of growth with the help of two strategic acquisitions. He and his management team continue to grow with the company. Similarly, Michael Cowpland has been able to manage Corel's growth from the start by concentrating on a few products with short life cycles.

As Gandalf grew, so did the number of products. It became increasingly expensive to maintain the more than 20 products that were developed from its original communications modem. The growth of product lines and the absorption of an acquisition forced some rationalization of activities. Something had to give. This was needed so that the firm could maintain a leading presence as a provider of communications networking equipment. One of the founders, Des Cunningham, had to come out of retirement more than once to guide the firm through difficult patches.

Mitel in its heyday was so entrepreneurial that it was frequently asked who was running the company. The firm overextended itself. British Telecom became the majority owner and brought in new management. Later another owner, Schroders Ventures, stepped in and further management changes ensued.

As these examples show, managing growth is a major challenge. The top officers in a company must accept the transition from entrepreneurial to professional management. Acceptance can be very difficult for a founder who has been involved in every aspect of the development of a firm and who cannot evolve with its growth.

Many firms opt to grow through alliances and acquisitions. Such a strategy has to be carefully managed. Firms usually underestimate the time, effort and financial resources that are required to successfully manage an alliance or absorb an acquisition. A key barrier to growth is insufficient working capital. In the steep part of the S curve, the rule of thumb for high-technology companies is that $1 of working capital is needed to support $1 of sales. At this stage, firms that do not have "deep pockets" like Corel and Newbridge are vulnerable to takeover or to failure.

Maturity and Renewal

For some 20 years ISE demonstrated that it was, and still is, a major player in the undersea unmanned vehicle market. That market is now a mature market. To get on another S curve and to renew itself, ISE is now attempting to apply its technology to space applications.

Instantel presents a similar situation, although it is still in its early development stage. It was very successful in penetrating its original market. However, the quarry and mining markets are relatively limited and if the firm wished to grow rapidly it would need to get into other markets with an equivalent product. So it opted to buy two new technologies for tracking and detection devices, on which it is pinning its hopes of growing at a rate of 50 percent per year. Firms that realize that their particular S curve could be limited do jump into other areas to maintain growth. The S curve then becomes the sum of a series of small S curves.

Cognos, on the other hand, has recognized that it is clearly in the mature phase with its existing product line. The firm is now shifting into the desktop and client/server market by moving its resources aggressively to this new area. At Gandalf, a new CEO is putting his mark on the firm by forcing the rationalization of old product lines while focusing on a new S curve with a strategy to penetrate the remote access market. With the market for PBX products shrinking, Mitel sees its future in the interface between tele-phony and computers.

Firms use a variety of strategies to move to the next S curve. They range from introducing new innovative products, to management changes, to new ownership. At maturity a firm cannot afford to sit back in its market niche. To ensure that it remains competitive it must renew itself. The renewal process can be difficult and can even fail. However, if renewal is not attempted, the firm will surely fail. Technology is moving too fast to permit firms to relax.

Conclusion

The dynamics of the development and growth of technology-intensive firms differ depending on where they are situated on the S curve at any given time. Firms get to a spot on the S curve at different times and with differing levels of effort. For example, Vortek is still looking for a way to climb the S curve after 20 years of effort, while Dynapro has experienced steady growth almost from the beginning.

Coping with the S Curve 221

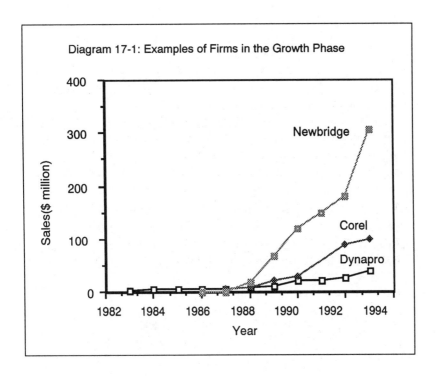

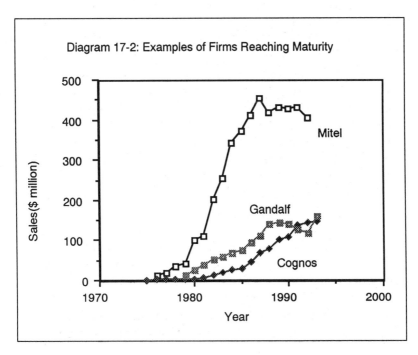

Newbridge and Corel, on the other hand, have seen spectacular growth, becoming $100 million companies in only five and eight years respectively (Diagram 17-1).

The dynamics of firm development centre on five interrelated elements: technology, products, markets, management and financing, the mix of which depends on a firm's position on the S curve. At the bottom of the S curve the need for leading-edge products and markets as well as financing from all possible available private and public sources dominate. In the growth phase, professional management and substantial financing to sustain growth, which is obtained primarily from the market place, are critical. At maturity, with the leveling-off of sales, there is a need to have the technology to develop the products and the management structure for a new round of growth on the next S curve. This is what Gandalf, Mitel and Cognos are currently going through, for example (Diagram 17-2).

18

The Context for Technological Innovation in Canada

It has been said that Canada is "the most developed of the developing countries," or "the largest of the smaller industrialized countries." Such descriptions make Canada some kind of "halfway house" between two worlds. This is where Canada finds itself today in the world of high tech. Because we have been successful at natural resource development, we did not have to live by our wits. We simply dug the stuff out of the ground and exported it, with everyone from the miner to the exporter making money in the process.

But the world has changed. Other countries are finding richer deposits of natural resources at a time when price and demand for these resources are eroding. So we are beginning to look elsewhere for new sources of wealth creation. We are finding that knowledge-based industries offer new venues for economic growth. If we are to develop these industries and move beyond the "halfway house" stage, we have to recognize that the following supporting elements are essential to growth.

People Are the Key

The major asset of a technology firm is its skilled people. This asset, which walks off the firm's premises every night, is highly mobile. Because of the risks involved, it is said that the technology business is a young person's business. Young people do not usually have many commitments. As Karl Brakhaus noted, "I didn't have kids, no one was married, there was no house, no mortgage — basically, we just didn't have a lot of overhead." Moreover, learning through failure is acceptable in the technology community, which means that skilled people can move from one situation to another with relative ease.

Because of the growth of knowledge-based industries, the demand for skilled people such as engineers, scientists and technologists is expected to increase in the future. Agencies such as the Natural Sciences and Engineering Research Council, the Canadian Council of Professional Engineers and the Technical Service Council all predict increased demand for technically skilled people. In 1950, some 40 percent of all jobs in North America were skilled jobs. It is estimated that by the year 2000 some 85 percent of jobs will be skilled.

There has been a significant decline in university enrolment in math, science and engineering programs in the 1980s. These careers were not perceived as being as attractive and remunerative as others such as medicine and business. In the past, shortages of technically skilled people were met through immigration, especially from Europe. Many of those immigrants went on to found very successful technology firms, such as Mitel, Newbridge and Corel. But with the post-war economic recovery, immigration from Western Europe has slowed down and is no longer the pool of technical people it once was. Eastern Europe has become a new source of skilled immigrants but, simultaneously, Canadians are also being attracted to jobs in other countries. The University of Waterloo, for example, has become an important source of computer science graduates for Microsoft in the US.

Whether there will be major shortages of technically skilled people in the future remains unclear. But it is clear that such individuals will migrate to where they are appreciated. It is up to Canadian policy makers to ensure that Canada remains an attractive environment for the skilled people that will be needed to develop the knowledge-based industries of the future.

Government Financial Support Is Essential

As indicated in the case studies, government financial support is very much needed, especially in the start-up phase of a firm where the traditional investment community fears to tread. Few investors accustomed to financing capital-intensive projects are comfortable financing people-intensive projects. Governments in industrialized countries have therefore stepped in to provide various forms of support to knowledge-intensive firms in their jurisdictions. These include tax incentives, grants, loans, contracts and even equity participation.

The Uruguay Round of the General Agreement on Tariffs and Trade (GATT) recognized that government research and development subsidies are legitimate. In a world economy whose fastest growing segments are based on technology, governments have the go-ahead to subsidize technology development. This is especially important for technology firms because they usually spend more than 10 percent of their revenues on R&D, a sizable expenditure.

In Canada, instruments for R&D support are found at both the federal and provincial levels. Canada's tax treatment for R&D is one of the most favourable in the world. A number of proven and well-appreciated programs, such as the Industrial Research Assistance Program (IRAP), are also in place. Overall however, Canada has one of the lowest levels of government spending on R&D in industry (grants and contracts) as a percentage of GDP among OECD countries. This is of concern if Canada is to move aggressively to develop knowledge-intensive industries which, as we have seen through the case studies, require not only the direct financial support of government, but also that government be a demanding buyer of new sophisticated products, which is the way that Gandalf was launched.

Moreover, government will have to streamline its procedures if it is to keep pace with the needs of fast-moving technology firms that do not have the time, resources and patience to go through bureaucratic labyrinths. Some programs, such as IRAP, with its 250 field officers across Canada, are good examples of how support programs should be delivered. Private sector vehicles, such as industry associations, that are closer to their members than government should also be used.

Knowledge-Based Industries Tend to Cluster

Knowledge-based firms tend to form industrial clusters in localities that have well-developed scientific and technological infrastructures which include universities and colleges, public and private research institutions and specialized facilities (such as nuclear accelerators, research parks, incubators). An agreeable social and physical environment is an added attraction for "footloose" skilled people and for retired experienced individuals who can act as mentors to young entrepreneurs.

Clustering is not surprising because innovation is a local phenomenon in that the innovator requires ready access to technical, financing and supplier support. Investors also prefer to invest locally so that they can interact closely with the firms they invest in.

As illustrated through the case studies, Vancouver and Ottawa are two such localities. Vancouver has two universities — the University of British Columbia and Simon Fraser University — as well as the British Columbia Institute of Technology, Discovery Park and the TRIUMF nuclear accelerator, among other institutions. Moreover, the city is situated in an enviable physical setting.

The Ottawa area also has two universities — the University of Ottawa and Carleton University — as well as Algonquin College and a relatively large concentration of government laboratories with their specialized facilities, such as the National Research Council's wind tunnel and the Communications Research Centre's satellite testing facility. The contribution of government laboratories to industrial development in the Ottawa area can be appreciated from Diagram 18-1. Bell-Northern Research also has a major research laboratory in the area. In the 1980s the Ottawa-Carleton Research Institute (OCRI) was created and given the mandate to further develop the

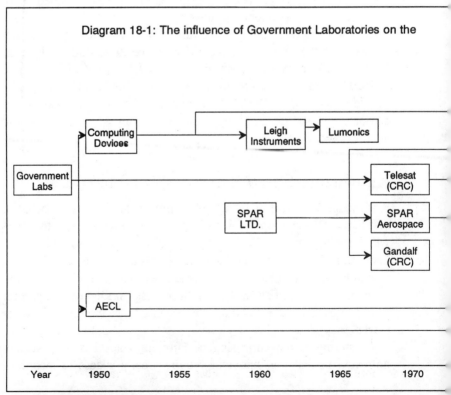

Diagram 18-1: The influence of Government Laboratories on the

Source: Doyletech Corp.

region's technical infrastructure and be the "glue" among the key elements of the infrastructure and the local high-tech community, which now has more than 400 firms. While Montreal and Toronto are major technical clusters, smaller ones include Sherbrooke, Kitchener-Waterloo, Winnipeg, Saskatoon, Calgary, Edmonton and Victoria.

Similar technology-oriented local clusters are emerging around the world. There are some 50 in the US, with Silicon Valley being the best known. Western Europe has a similar number with southern localities such as Grenoble, Montpellier and Sophia-Antipolis, France, emerging as attractive regions. Japan passed the technopolis law in 1983 to set up 26 technology clusters across the country in the next 20 years, to relieve pressure on the Tokyo area.

The emergence of local technology clusters has municipal and regional authorities around the world competing to attract knowledge-based investment to their communities. To make themselves as attractive as possible, they invest in the type of technical infrastructure that will appeal to industry. This includes setting in place high-

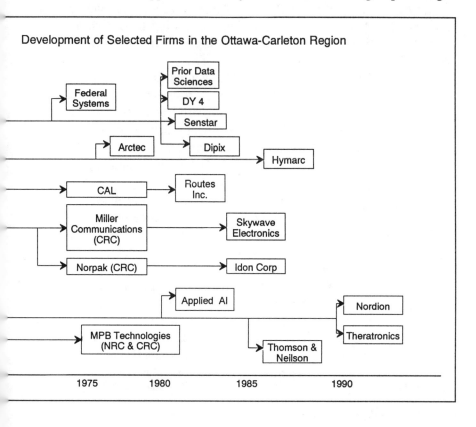

speed communications networks, lobbying for government laboratories and specialized educational and technical facilities, not to mention direct air links to major commercial centres.

The local-clustering phenomenon is challenging governments to develop the programs needed to serve these knowledge-based centres or technopolies. This presents a major challenge to local economic development officials in Canada whose tradition has been to sell or lease serviced land, rather than to develop the technical underpinnings needed to support industrial development.

The globalization of the world economy is being driven by the removal of economic barriers between countries. In such an environment the local level increasingly becomes the crucible where innovation occurs.

The World Is the Stage

As demonstrated through the case studies, knowledge-based firms have to export to survive. It is not unusual for these firms to export 80 to 90 percent of their production. While the US is the favoured market, Canadian firms are increasingly exporting to Europe and Asia. Technology-based exports are growing and currently represent about 15 percent of Canada's exports.

To facilitate exports, firms increasingly get into alliances with like firms having a presence in other markets. An example is the long-term relationship that Dynapro established with Allen-Bradley. Marketing, distribution and product development alliances are now an accepted way of doing business internationally. Even smaller firms, such as Instantel, find it advantageous to link up with similar firms in other jurisdictions.

Traditionally exports led investment because it only made economic sense to establish a presence in a market once the level of sales warranted it. Now investment increasingly leads trade; a presence in a market is becoming a precondition for selling in that market. Some countries, such as France, insist on it.

Larger firms have been setting up alliances to help them penetrate markets for some time. The federal government is now encouraging the formation of international alliances between smaller firms through a brokerage program that utilizes experts in knowledge-based industries, such as telecommunications, biotechnology and software, as go-betweens to match up like firms. This is an innovative approach that complements other export-support programs.

As Canadian firms move beyond the US market into the Asian, European and Latin American markets, they have to be more sensitive to different cultures and languages. While the language of international commerce is still English, sensitivity to the milieu in which the firm operates is not only politically correct, it is good business. For example, by translating CorelDraw into 22 languages, Corel found that it could double or triple sales in each of those markets.

Summing Up

The new innovators are driving the creation of a knowledge-based economy in Canada. This economic transformation is still in its early days. The paradigm shift needed to fully appreciate the implications of an economy based on the creativity of skilled people rather than one based on capital-intensive resource extraction has only just begun to take place.

By the late 1980s, Canadian technology-intensive exports were about 15 percent of total exports, up from 11 percent in 1971. The paradigm is shifting slowly. However, the proportion of technology-intensive exports was about 40 percent in the United States, 30 percent in Japan and 25 percent in the newly industrializing countries of Asia. Canada is in a catch-up mode relative to its competitors.

Similarly, using the classic OECD indicator of the ability to innovate, that is the Gross Expenditures on Research and Development (GERD) as a percentage of Gross Domestic Product (GDP), Canada comes in at 1.5 percent, well below the 2.5+ percent of Sweden, Japan, Switzerland, Germany and the United States. In fact, Canada's GERD/GDP ratio is at about the same level as that of Malaysia and is lower than those of non-OECD countries like Taiwan (1.7 percent) and Korea (2 percent).

Whatever indicator of knowledge or technology intensiveness is used, Canada will still fall behind other major, and not so major, countries. Apologists for this situation argue that Canada's resource-based economy does not require as much R&D and related investment as do the economies of countries more involved in manufacturing. The culture that relies on this type of reasoning is not particularly supportive of the development of knowledge-based industries which have high R&D expenditures and need financial support.

There exists a substantial gap between the political rhetoric which calls for a new knowledge-based economy and the reality which is well anchored in a risk-averse status quo. Knowledge-based firms do not operate in a particularly friendly environment in Canada. There is not much help available to lower the risks faced in starting and developing knowledge-based firms. As a result, it is surprising that some of these companies do as well as they do. One can only imagine the possibilities if Canada's leaders ever decided wholeheartedly to support the knowledge-based economy.

List of Acronyms

A-B	Allen Bradley
ATM	Asynchronous Transfer Mode
AUV	Autonomous Underwater Vehicle
BCCA	British Columbia Cancer Association
BNR	Bell Northern Research
BPD	benzoporphyrin derivative
BT	British Telecom
CBC	Canadian Broadcasting Corporation
CD	Compact Disk
CD-ROM	Compact Disk-Read Only Memory
CRTC	Canadian Radio and Telecommunications Commission
CSA	Canadian Space Agency
DIPP	Defence Industry Productivity Program
DND	Department of National Defence (Canada)
FDA	Food and Drug Administration (USA)
GATT	General Agreement on Tariffs and Trade
GDP	Gross Domestic Product
HP	Hewlett-Packard
HPB	Health Protection Branch (Canada)
IP	Intellectual Property
IRAP	Industrial Research Assistance Program
ISDN	Integrated Services Digital Network
ISTC	Industry, Science and Technology Canada
LAN	Local Area Network
MIT	Massachusetts Institute of Technology
NASA	National Aeronautics and Space Administration (USA)
NASDAQ	National Association of Securities Dealers Automated Quotations
NEC	Nippon Electric Corporation
NRC	National Research Council (Canada)
NSTF	National Solar Testing Facility

OECD	Organization for Economic Cooperation and Development
OEM	Original Equipment Manufacturer
PBX	Private Branch Exchange
PC	Personal Computer
PRAI	Project Research Applied to Industry
R&D	Research and Development
RF	Radio Frequency
ROV	Remotely Operated Vehicle
S&D	Search and Develop
SCSI	Small Computer System Interface
TSE	Toronto Stock Exchange
UBC	University of British Columbia
VME	Versa Modular Europe
WAN	Wide Area Network
WDF	Western Economic Diversification Fund

Bibliography

Arcangeli, F., P. David, and G. Dosi. *Frontiers in Technology Diffusion*. Toronto: Oxford University Press, 1991.

Brown, J.J. *Ideas in Exile*. Toronto: McClelland and Stewart, 1967.

Brown, W., and R. Rothwell, editors. *Entrepreneurship and Technology*. London: Harlow U.K. Longman, 1986.

Cantwell, J. *Technological Innovation and Multinational Corporations*. Oxford: Basil Blackwell, 1989.

Davies, S. *The Diffusion of Process Innovations*. Cambridge and New York: Cambridge University Press, 1979.

De Jordy, H. *Blueprint for a Country Turnaround*. N.p.: CdC International Press Ltd., 1992.

Dosi, G., et al. *Technical Change and Economic Theory*. London: Pinter Publishers, 1988.

Doyle, D. *Making Technology Happen*. 3rd edition. Kanata, Ontario: DoyleTech Corp., 1992.

Dufour, P., and J. de la Mothe, editors. *Science and Technology in Canada*. London: Longman, 1993.

Foster, R.N. *Innovation: The Attacker's Advantage*. New York: Summit Books, 1986.

Freeman, C. *The Economics of Industrial Innovation*. London: Pinter Publishers, 1982.

Freeman, C., and L. Soete, editors. *New Explorations in the Economics of Technical Change*. London: Pinter Publishers, 1990.

Gibbons, M., and R. Voyer. *A Technology Assessment System*. Ottawa: Science Council of Canada, 1974.

Hammer J., and J. Champy. *Re-engineering the Corporation*. New York: Harper Business, 1993.

Handy, C. *The Age of Paradox*. Boston: Harvard Business School Press, 1994.

Hippel, E. von, *The Sources of Innovation*. New York: Oxford University Press, 1988.

Lamontagne, M. *Business Cycles in Canada.* Toronto: James Lorimer and Co., 1984.

Lundvall, B.A. *National Systems of Innovation.* London: Pinter Publishers, 1992.

Mansfield, E. *Industrial Research and Technological Innovation.* New York: W.W. Norton, 1968.

Oakly, R., R. Rothwell, and S. Cooper. *Management of Innovation in High Technology Small Firms.* London: Pinter Publishers, 1988.

OECD. *Technology and the Economy.* Paris, 1992.

Peters, T. *Liberation Management.* New York: Alfred A. Knopf, 1992.

Porter, M.E. *The Competitive Advantage of Nations.* London: MacMillan, 1990.

Roberts, E.B. *Entrepreneurs in High Technology.* New York: Oxford University Press, 1991.

Schofield, B.T., and R. Thomson. *Technological Change and Innovation in Canada.* Government of Canada, 1988.

Smilor, R.W., et al. *Creating the Technopolis.* London: Ballinger Publishing Co., 1989.

Voyer, R., and M. Murphy. *Global 2000: Canada, A View of Canadian Economic Development Prospects, Resources and the Environment.* Oxford: Pergamon Press, 1984.

Index

Access automated pap testing, 46, 52, 53
Adobe Systems Inc., 117
Albach, Gary, 21, 22, 23, 26, 27-28
Aldus Corp., 117
American Cyanamid Co., 36, 38
Angus, Bob, 139
Annatech, 23
asynchronous transfer mode (ATM), 125
Atlas, 76
automatic battery cycler, 63-64
autonomous underwater vehicles (AUVs), 202, 205

Babbage, Charles, 7
Baxter Healthcare Corp., 39
Baylor Research Institute, 42
BC Research, 24
BC Science Council, 60, 88
Bell Northern Research (BNR), 58, 71
benzoporphyrin derivatives (BPDs), 36, 41
Black, Terry, 96, 100
Boeing, 25
Brakenhaus, Karl, 139-150
British Columbia Cancer Association (BCCA), 45
British Telecom, 109; purchase of Mitel, 175
Brown, J.J., 1-2
Brown, John, 32
bureaucracy, 106, 118
Burgess, Wendy, 186, 187, 191, 193, 194-195, 198, 199-200
business plan, 15-16, 55

Camm, David, 20, 21, 22-25, 26, 28, 29
Canada: financing problem in, 16; foreign ownership in, 4; natural resource exploitation, 4; technology-based industries, 2

Canada Centre for Remote Sensing, 88
Canada Export Award, 125
Canadian firms, 5
Canadian Shipbuilding & Engineering Ltd, 205-206
Canadian Space Agency (CSA), 202
capital, 54; and small firms, 4, 209-210. *See also* venture capitalists
CBC Pensions, 98, 99
CD-ROMS, 110
Cecil Green award, 143
climbing the S Curve, 218-220
Clohessy, Kim, 96, 100
Cluster Module Controller, 63
clustering, 225-228
Cognos Inc.: alliances, 159; artificial intelligence research, 156; competition, 157, 166; development of Quiz, 154;
financing, 155, 156; first products, 153-155; geographical markets, 160; and information systems, 157; and open systems, 159, 161; R&D, 161; rapid growth, 156; shift to programs for non-experts, 158-159
commercialization. *See* technological innovation
computer, development of electronic, 7-8
computerized dispatch system, 190
CONTROL-Cision Integrated Control Software, 63
convenience relationships, 55-56
copyrights, 12
Corel Corp.: competition, 117-118; desktop publishing systems, 110-111; international market, 114; management technique, 119; market niche, 113, 115; market share, 116; marketing, 113, 119; R&D, 113; strategic alliance, 116; working capital, 114

236 The New Innovators

CorelDraw, sales, 110, 112
CorelSCSI, 112, 116
CorelShow Runtime Player, 112
Cowpland, Michael, 109-120, 170, 173-174, 175, 179, 181
Creo Products Inc.: management strategy, 92-94; market, 89; optical tape recorders, 85-86, 87, 91; photoplotter, 86, 88, 91; R&D, 90, 92; strategic alliances, 86
Csathy, Thomas, 156, 157, 158, 162, 167
Cunningham, Desmond, 186, 187, 188, 189, 191, 192, 193
Cyanamid Canada, Inc., 37

Dainippon Screen Manufacturing Co. Ltd., 86, 88
Dallas Instruments, 75
Defence Industry Productivity Program (DIPP), 61, 104, 178, 207
Department of Industry, Science and Technology, 206
Department of National Defence, 88
Department of Regional Economic Expansion, 178
deregulation of phone industry, 126, 171, 173, 186
Diamond Taxicab Association, 190
Dietrich, Rob, 171, 177, 178, 180-181, 182, 183
discovery, defined, 6
Dolphin, David, 32
Dool, Garry, 95-107
Dowty Group PLC, 192
Doyle, Denzil, 71
DY 4 Systems Inc.: competition, 97, 103-104; Danish Navy contract, 98; entrepreneurial spirit, 106-107; geographical markets, 100; Indonesian catastrophe, 99; market niche, 98-99, 101; market share, 102-103; quality assurance, 106; R&D, 104-105; restricted customer base, 102
Dynapro Systems Inc.: Allen-Brady alliance, 140-141; management style, 147; and manufacturing locally, 145-146; market share, 147; purchase of TFP, 144; risk-taking, 149-150; sales, 142; touchscreen computers, 143-144

Eaton Corp., 25
Edison, Thomas, 2
Endoscan, 45
Energy Conservation Systems (ECS), 205
ENIAC computer, 8
Enterprise Development Program, 173
entrepreneur, 17, 33
European Space Agency, 88

Fairclough, John, 138
Federal Business Development Bank, 20, 23, 88
financing innovation, 16
first mover advantages, 41
first products, 216-218
Flight Dynamics Laboratory, 25
Food and Drug Administration (FDA), 52
Foran, William, 37, 38, 40, 43
free trade, 92

Gandalf Technologies Inc.; and government subsidies, 195; interest in Case Group PLC, 191-192; markets, 195; merger with Infotron, 192-193; remote access, 197-198; sales, 188-189, 190; shares, 189;
Gelbart, Dan, 85
GeoSonics, 75
Gibbs, Donald, 179
Glenister, Peter, 151
government funding: Cowpland's view on, 120; Mackie's view on, 127-128
government support, 4-5, 23, 104, 224-225; Creo Productions, 88; Mitel, 178; Potter's view on, 160; for start-up companies, 38-39, 61; of Xillix, 48
GRAFIX terminal, 140
Griffiths, Anthony, 175, 176, 177
Gross Expenditures on Research and Development (GERD), 229
Guandong Enterprises Corp, 35
Gunn, William, 138, 139

Hamamatsu, 53
Hanssman, Michael, 58
Health Protection Branch (HPB), 40
Hedges, Brian, 193, 194
Honeywell, 86

IBM, 8, 97; alliance with Rolm, 174, 181-182
ICI Imagedata, 86
Industria Farmaceutica Cosmetica Italiana, 35
Industrial Research Assistance Program (IRAP), 23, 61, 78, 88, 104
Industry, Science and Technology Canada (ISTC), 48, 172
innovation: defined, 6; process, 7, 9-11. *See also* technological innovation
InQuizitive, 158
Instantel Inc.: competition, 75, 81; funding, 78; growth, 73, 82; R&D, 74, 79, 80, 81; revenues, 78; seismograph, 74-75; tracking devices, 77-78
Integral Services Digital Network (ISDN), 187, 192
intellectual property, 11-13
International Submarine Engineering (ISE): AUVs, 205; financing, 203; and lack of capital, 209-210; and market creation, 212; R&D, 209; Robotic Refuelling system, 206-207; ROVs, 204; and Spar, 207
International Transportation Systems (STS), 205
invention, defined, 6
Isherwood, Barclay, 49

Jackson, Erin, 207
Jaggi, Bruno, 45, 48
Jarvis, John, 175, 176
John Fluke Manufacturing Company Ltd., 143

Kodak, 53

Laser Zentrum Hannover, 26
Leibniz, Gottfried Whilhelm, 7
Levy, Julia, 32, 33
Life fluorescence endoscope, 46, 52
local network controllers, 124

McFarlane, Jim, 201, 203, 204, 206, 207, 208, 209, 210, 211, 212
Mackie, Jim, 121, 124, 125, 126, 127, 128, 130, 131, 132, 133, 134, 135, 136
McMahon, Richard, 58-69
Main, David, 33, 35, 37, 38-39, 42, 43-44
management discipline, 54
managing innovation, 17-18
market analysis, 14
market forecasting, 105
market-pull innovation, 7
markets, creating, 28
Martin, Brian, 70-83
Matthews, Terrence, 109, 170, 173-174, 175
maturity and renewal, 220
medical market, 42, 51-52
Micom AES word-processor users, 109
Micrografx, 117-118
MicroImager, 46, 47, 49, 53
Microscan, 45
Microscan Imaging and Instrumentation Incorporated, 46
Microsystems International, 123
Millard, John, 177, 182
Miller, Jim, 32, 35, 37
Ministry of Economic Development, 48
Minns, Bob, 151, 152, 153-155, 160, 161, 162, 163, 165
Mitel Corp., 109; declining market for PBX, 176-177, 180-181; and deregulation of telephone industry, 171, 173; financial loses, 174-177; financing, 170; PBX revenues, 178; R&D, 178; and SX2000, 172-174, 181
Mitel Corp. British Telecom management of, 184-185
Motorola Inc., 103
multiplexers, 124

NASA, 24
National Research Council (NRC) grant, 20, 60, 61, 88
National Solar Testing Facility (NSTF), 23
Newbridge Networks Corp.: customers, 134; deregulation of phone industry,

238 The New Innovators

126-127; employees, 135; expansion, 125; financing, 123-124; geographical markets, 128; image problem, 129-130; management style, 135-136; R&D, 128-129; shares, 125, 126
Nodwell, Roy, 20, 23
Noranda Enterprise Ltd., 98, 99, 155

OCRA Communications, 190
Ontario Centre for Resource and Mining Technology, 78
Ontario Technology Fund, 159
Open Systems Interconnect (OSI), 187
optical tape recorders. *See* Creo Products Inc.
Original Equipment Manufacturer (OEM), 154
Osadca, Danny, 99, 100

Pacic, Branko, 45, 48
PACX, 187, 188
PAL-68000 computer system, 60
Papows, Jeffrey, 158, 159, 167
Parsons, Robert, 59
partnerships, 42
patents, 12
Patterson, Colin, 186, 187, 188, 189
Phillips, Anthony, 32
PhotoCD, 112
photodynamic therapy. *See* Photofrin; benzoporphyrin derivatives
Photofrin, 32, 33-34, 36, 38, 39, 40, 41
Photomedica Inc., 36
Photometrics, 53
photon furnace, 25
photoplotter. *See* Creo Products Inc., 86, 88
Potter, Michael, 151-153, 155, 156, 157, 158, 159, 160, 161, 162, 163-164, 165, 166-168
PowerHouse, 155, 158
product migration, 105
product migration strategy, 15-16
Project Research Applied to Industry (PRAI) grant, 23
proprietary technology, 2

Quadra Logic Technologies Inc. (QLT): business plan, 39-40; financing of, 34, 36-37; first mover advantages, 41; low cost of Photofrin, 43; products, 34-37; professional management of, 43; R&D, 38; rat scheme, 33
Quasar Systems Ltd., 151
Quick, 145-155
Quiz: development of, 154; sales, 152, 154

Radstone Teck, 103-104
rapid growth phase, managing, 212-213
Raytheon Canada Ltd., 99
Redifacts Advanced Manufacturing Aids Ltd., 189-190
remotely operated undersea vehicles (ROVs), 202, 205
research and development (R & D), 2; in Canada, 5; and large firms, 3, 4
Richards, Steve, 96, 100
Rockwell Systems Australia Pty., 99
Routtenberg, Michael, 45, 49
Rushforth, Alan, 151
Russell, Bertrand, 7

S curve, 14, 15, 16, 17, 54; stages, 9-11, 214-222
Schroder Ventures, purchase of Mitel, 177
search and develop strateg(ies) (S & D), 2-3, 13
Semi-Conductor Research Corporation, 68
Shanghai Institute of Planned Parenthood, 34
short sellers, 130
Simple Network Management Protocol (SNMP), 187
skilled people, 223-224
Software Publishing Inc., 117
Sommerer, Peter, 131
Spar Aerospace Ltd., 207
Spencer, Ken, 85-94
"sputtering," 58
start-up, 214-216
STD bus, 96
strategic alliances, 86
subcontractors, 213
suppliers, as stakeholders, 55

Index 239

Sutcliffe, David, 45, 47-48, 49-51, 52-53, 54-56

TC-111 Tool Controller, 63
technological innovation, 1-5; and large firms, 19; and small firms, 4
technology, defined, 6
technology transfer, 13, 68
technology transfer company. *See* Vortek Industries Ltd.
technology-based exports, 228-229
technology-push innovation, 7
Techware II Process Equipment Controller, 63
Techware Systems: business strategy, 67-68; financing of, 60-61; markets, 66; and R&D, 62; and recession, 64; sales, 61; start-up, 59-60; and threatened foreclosure, 65; uniqueness, 63, 67
Thin Film Production (TFP), 144
third-party control system, 67
Total Alert, 77
trade secrets, 13
trademarks, 12-13

UBC, 47, 48, 49, 85, 139, 151; and McMahon, 58-59; and QLT, 33, 34; and Vortek Industries, 20, 23
United States International Trade Commission, 75
University of Waterloo, 224
US Commerce Department, 75

Vassiliades, Thomas, 194, 200
Ventura Software Inc., 116
venture capitalists, 16
Vibra-Tech, 76
VME open architecture computer systems, 96
Vortek Industries Ltd.: business plan, 30-31; technology transfer company, 21-28
Vortek lamp, 20-22, 24, 25, 26-27

W.H. Brady Co., 144
Walt Disney Studios, 21
Western Economic Diversification Fund (WDF), 48, 61, 142, 143
Whitehead, Alfred North, 7

Woodward, Henry, 1
Wright-Patterson Air force Base, 25, 206

X-30 "space plane", 26
Xillix Technologies Corp.: financing of, 48, 51; potential market, 50; products, 46; professional management, 49, 54; R&D, 50; rapid growth phase, 54; reference sales, 50; strategic alliances, 53

Zambonini, Ron, 158, 159, 165, 167